Environmental Footprints and Eco-design of Products and Processes

Series Editor

Subramanian Senthilkannan Muthu, Green Story Inc., Toronto, Canada

Indexed by Scopus

This series aims to broadly cover all the aspects related to environmental assessment of products, development of environmental and ecological indicators and eco-design of various products and processes. Below are the areas fall under the aims and scope of this series, but not limited to: Environmental Life Cycle Assessment; Social Life Cycle Assessment; Organizational and Product Carbon Footprints; Ecological, Energy and Water Footprints; Life cycle costing; Environmental and sustainable indicators; Environmental impact assessment methods and tools; Eco-design (sustainable design) aspects and tools; Biodegradation studies; Recycling; Solid waste management; Environmental and social audits; Green Purchasing and tools; Product environmental footprints; Environmental management standards and regulations; Eco-labels; Green Claims and green washing; Assessment of sustainability aspects.

Subramanian Senthilkannan Muthu
Editor

Ecological Footprint of Industrial Spaces and Processes

Springer

Editor
Subramanian Senthilkannan Muthu
Green Story Inc.
Toronto, ON, Canada

ISSN 2345-7651 ISSN 2345-766X (electronic)
Environmental Footprints and Eco-design of Products and Processes
ISBN 978-3-031-69046-4 ISBN 978-3-031-69047-1 (eBook)
https://doi.org/10.1007/978-3-031-69047-1

This Springer imprint is published by the registered company Springer Nature Switzerland AG
The registered company address is: Gewerbestrasse 11, 6330 Cham, Switzerland

This book is dedicated to:
The lotus feet of my beloved Lord
Pazhaniandavar
My beloved late Father
My beloved Mother
My beloved Wife Karpagam and Daughters—
Anu and Karthika
My beloved Brother
Last but not least
To everyone working in various industrial
spaces to make the world "SUSTAINABLE"

Contents

Introduction to the Book—Ecological Footprint of Industrial Spaces and Processes

Subramanian Senthilkannan Muthu

With the overshooting consumption of resources from nature, it's highly imperative and emergent to measure the dependence of human activities on the nature, rather on the natural resources. On the line of basic definition of sustainability, we have to clearly understand that how much of the environment or earth's resources are needed to sustain any life style on our mother earth. Ecological footprint is one of the damn important tools which can be deployed to the measure of impacts of an individual or community on the environment (on our earth) and it can be expressed as the amount of resources consumed.

The concept of ecological footprints was introduced in 1990 for the evaluation of environmental impacts associated with human activities by Prof. Wackernagel and Rees [1]. In fact, this precedes the introduction of well known, carbon footprint methodology. Ecological footprints do quantify the rate at which resources are consumed and wastes are generated.

This book deals around the concept of Ecological footprint and how this important measure is quantified across various industrial spaces. Beginning from explaining the concept of ecological footprint and its implications on sustainability, followed by the detailed assessment of ecological footprint in five distinct industrial spaces. Each industrial space is quite different in terms of their ecological issues and hence the quantification of ecological footprints and hot-spots arising out of each sector are quite unique as well. This is worth exploring and this is detailed in this book through five informative chapters.

The second chapter deals around the ecological footprint in textiles and clothing sector. The Ecological Footprint analysis of the textile industry may help policymakers make decisions to promote more sustainable practices and contribute to developing environmentally friendly textile production methods. This study

S. S. Muthu (✉)
Green Story Inc., Toronto, ON, Canada

S. S. Muthu (ed.), *Ecological Footprint of Industrial Spaces and Processes*,
Environmental Footprints and Eco-design of Products and Processes,
https://doi.org/10.1007/978-3-031-69047-1_1

mentioned in the second chapter explores the environmental impact of the cloth (fabric) manufacturing process. This study assesses attempts to qunatify the resource consumption, manpower requirements and overall environmental impact of textile (fabric) production in India. From the results, it is found that the Ecological Footprint of cotton fabric and synthetic fabric production from their raw materials is 16.64–18.17 gha/ton and 10.2–11.87 gha/ton; respectively. The Ecological Footprint of cotton fibre production is more than polyester fibre production. This is because cotton cultivation needs huge cropland (bioproductive land). The environmental impact of natural/man-made fabric production is significantly higher and it can be reduced by simply reducing the demand for textile products made from these fibres. Authors of this chapter believe that the reported study is an eye-opener in the field of the textile industry and adopting simple steps (such as reducing, recycling and reuse of textile products) may help to achieve the United Nations Sustainable Development (Goal No 12—Responsible Consumption and Production). As per the authors, findings arising from this research reported in this chapter, provide valuable insights into the sustainability challenges posed by power loom textile manufacturing, and suggest potential strategies for reducing its environmental impact.

The third chapter in this book talks about the analysis of the Ecological Footprint of Dry Chemical Silage Derived from Red Tilapia (Oreochromis spp.) Viscera. It's evident that there is a global increase in fish production, mainly driven by growing population and consumer interest. This has brought forth many environmental challenges, particularly in the form of waste generated by this industry. Specifically, the disposal of viscera has emerged as a significant contributor to negative environmental impacts and authors wish to find the solutions for the same through the study reported in this book as the third chapter. This study focuses on assessing the environmental repercussions associated with obtaining dry chemical silage from the viscera of red tilapia (Oreochromis spp.), employing the ecological footprint methodology as a comprehensive indicator of sustainability. The findings from this study reveal that the conversion of fish viscera into dry chemical silage brings about a noteworthy reduction, approximately 30%, in carbon dioxide (CO_2) emissions compared to the alternative scenario where fresh viscera are simply discarded into shallow landfills. This reduction is a critical aspect of environmental sustainability, given the role of CO_2 in contributing to climate change. Moreover, the authors of this study identify specific actions that can be implemented to further enhance the sustainability of this process. For instance, incorporating measures like biogas production from waste and employing solar drying techniques for the final product can lead to an impressive mitigation of approximately 86% of the overall environmental impact. This emphasizes the importance of adopting innovative and eco-friendly practices in the production and waste management stages of the process. By the way of conclusion, the authors who have performed this research establishes that the production of dry chemical silage from red tilapia viscera, especially when utilizing alternative drying energy sources, is environmentally sustainable. Authors conclude that this insight is valuable not only for the aquaculture industry but also for broader discussions on sustainable practices within the context of food production and waste management.

The fourth chapter in this book is going to talk about the ecological footprint of cement sector. Authors of this chapter have performed and reported the Ecological Footprint and Economic Assessment of Ready-Mix Concrete Production. As mentioned in this chapter, cement industries constitute approximately 8% of global anthropogenic carbon dioxide emissions, including cement by-pass dust, making them a critical environmental concern due to their direct impact on human health and global warming. Typically, conventional concrete production utilises cement as a binding material. On the contrary, the study reported in this book as the fourth chapter addresses this critical issue by using Fly ash and GGBS (Ground granulated blast-furnace slag) to partly replace cement in M20, M25, and M30 grades concretes. The reported study evaluates the environmental impact of concrete production with supplementary cementitious material for normal concrete using the Ecological Footprint (EF) tool for ready-mixed concrete (RMC) plants. The reported and analysed results showed that the fly ash and GGBC can effectively replace cement in conventional concrete, with meet the target comparable compressive strength. The annual Ecological Footprint of the ready-mix concrete (RMC) plant is estimated as 919.05 gha, as reported in this assessment. The Ecological Footprint of M20, M25, M30 and M35 concretes are found to be 0.0268, 0.0272, 0.0284 and 0.0305 gha/m^3; respectively. The study suggested that the Ecological Footprint of materials used for concrete production has the highest among all other parameters like materials transportation, energy consumption, labour etc. Concrete production using the combination of fly ash and GGBS (as a partial replacement of OPC) has a lower environmental impact than conventional concrete (consisting of OPC) production. The authors of this chapter concluded that using Fly ash and GGBC in concrete production can be an innovative solution because it simultaneously solves the problem of solid waste disposal, carbon emissions, cost and environmental impact.

The fifth chapter is dedicated to quantifying the environmental footprint through an in-depth assessment of the environmental impact associated with the Production of California Red Earthworm Meal (CRWM) using the ecological footprint methodology. The comprehensive environmental impact assessment of Californian Red Worm meal (CRWM) production involves a laborious examination of various ecological parameters to gauge the sustainability of this agricultural practice. One crucial aspect of the analysis is the evaluation of resource utilization throughout the entire lifecycle of CRWM, encompassing stages such as raw material acquisition, processing, and distribution. By scrutinizing the resource inputs and outputs at each stage, the reported study resolves around to identify the potential areas for optimization, resource efficiency, and waste reduction. The process of obtaining Red Californian Earthworm Meal stands out as an environmentally beneficial practice, contributing to a substantial reduction of 84 kg of CO_2 emissions per kilogram of meal annually, as assessed and reported by the authors of this fifth chapter. This noteworthy reduction is primarily attributed to the intricate breeding and reproductive stages of the earthworms. The implementation of ecological footprint as a methodology for environmental assessment of the Californian Red Worm paste production process (Eisenia fetida) discussed in this chapter, allowed the authors to quantify the true environmental impact of utilizing waste from agro-food and livestock industries, identifying stages and input-output flows with the highest CO_2 emissions. As per the authors of this chapter, this is valuable for the implementation of new alternative technologies, achieving both technologically and environmentally sustainable processes. The authors have also added that additionally, vermiculture presents high feasibility as it generates two products with significant biological and economic value: vermicompost and worm paste, thereby reducing environmental impact through the utilization of agro-industrial waste.

Last, but not the least, the sixth chapter reports the Ecological Footprint of Electric Vehicle Charging Infrastructure. As per the authors of this chapter, the three main factors affecting the Ecological Footprint (EF) of electric vehicle (EV) charging stations are direct and indirect emissions (due to electricity consumption), labour and physical land required. Electricity is required for electric vehicle charging stations to recharge EV batteries, and the electricity source has a big impact on the environment. The EV charging station may reduce its Ecological Footprint by using a solar PV power system. As per the calculation results, the Ecological Footprint of EV charging is about 3.1×10^{-4} gha/kWh of battery charging. However, the proposed solar PV system by the authors in this chapter may reduce the environmental impact of battery charging to 1.45×10^{-4} gha/kWh. According to the authors, for an eco-friendly electric charging system, it is crucial to promote sustainable energy sources. Since using renewable energy sources is the key to minimizing the environmental implications of electric vehicle charging stations, it is also crucial to encourage consumers to adopt them. The authors have performed the economic assessment of EV charging station as well and reported the findings in this chapter. The authors have also listed the limitations of this study as well.

I take this opportunity to thank all the contributors for their earnest efforts to bring out this book successfully. I am sure readers of this book will find it very useful.

With best wishes,
Dr. Subramanian Senthilkannan Muthu

Reference

1. Wackernagel, M., & Rees, W. (1996). *Our ecological footprint: Reducing human impact on the earth*. New Society Publishers.

Ecological Footprint Assessment of the Textile Industry in India

Mahvi Malik Shahzad, Shaikh Parvez, Nusrat Shafeeque Ahmad, Dilawar Husain, Md Azhar, Sajid Naeem, and Ashok Kumar Pandey

Road Map of the Chapter

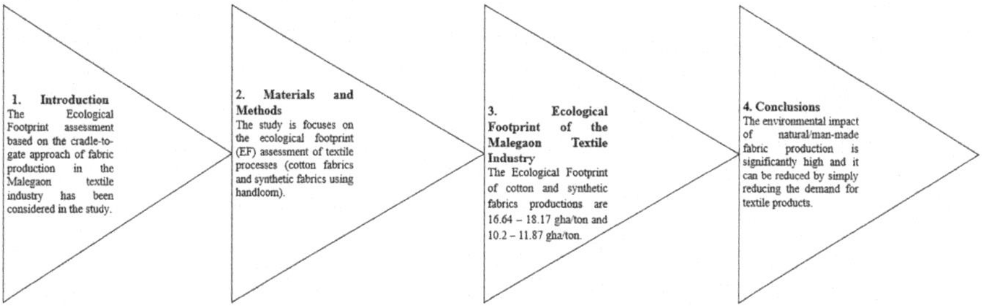

M. M. Shahzad · S. Parvez · D. Husain (✉) · M. Azhar
Department of Mechanical Engineering, Maulana Mukhtar Ahmad Nadvi Technical Campus, Malegaon, Maharashtra, India

N. S. Ahmad · S. Naeem
Department of Applied Sciences, Maulana Mukhtar Ahmad Nadvi Technical Campus, Malegaon, Maharashtra, India

A. K. Pandey
Training & Placement Officer Engineering, Government Polytechnic College, Shahdol, M.P, India

S. S. Muthu (ed.), *Ecological Footprint of Industrial Spaces and Processes*, Environmental Footprints and Eco-design of Products and Processes,
https://doi.org/10.1007/978-3-031-69047-1_2

1 Introduction

The clothing industry is a significant ecological imprint on the planet. Fabric consumption has risen exponentially in the past decade [1]. The textile industry is estimated to be responsible for 10% of the total global GHG emissions as well as it is responsible for about 20% of the total global clean water pollution [2]. The projected 2030 average emissions of fashion consumption for G20 countries are the highest in Australia (i.e., per capita 503 $kgCO_{2e}$/year), and the lowest in India (i.e., per capita 22 $kgCO_{2e}$/year). Therefore, an essential emissions reduction of up to 60% (per capita 128.7 $kgCO_{2e}$ by 2030) is needed to meet the target of the Paris Agreement [3]. Due to fast fashion demand, it is anticipated that this tendency will continue, with a 63% increase in fabric demand by 2030 [4].

Cotton has been used to manufacture fabrics since ancient times and is currently the most popular natural (fibres) material for textile production. Some other synthetic (fibres) materials are also used for textile production [1]. The Indian textile industry is anticipated to grow by 10% from 2019 to 2020, reaching 190 billion USD by 2025–26. With a 4.6% share of the global trade in textiles and apparel, India is ranked as the world's 3rd largest exporter in this sector. In 2022–23, fibre production in India amounted to 2.15 million tonnes, while yarn production reached 5185 million kgs during the same period. Natural fibres are considered the foundation of the Indian textile industry, which is set to expand from 138 to 195 billion USD by 2025. The textiles industry in India employs around 4.5 crore workers, including 35.22 lakh handloom workers nationwide [5].

Textile fibre types are classified into two major categories natural and manufactured (man-made) fibres. The detailed classification of textile fibres is shown in Fig. 1 [6]. Natural fibre, mainly cotton fibre requires large amounts of

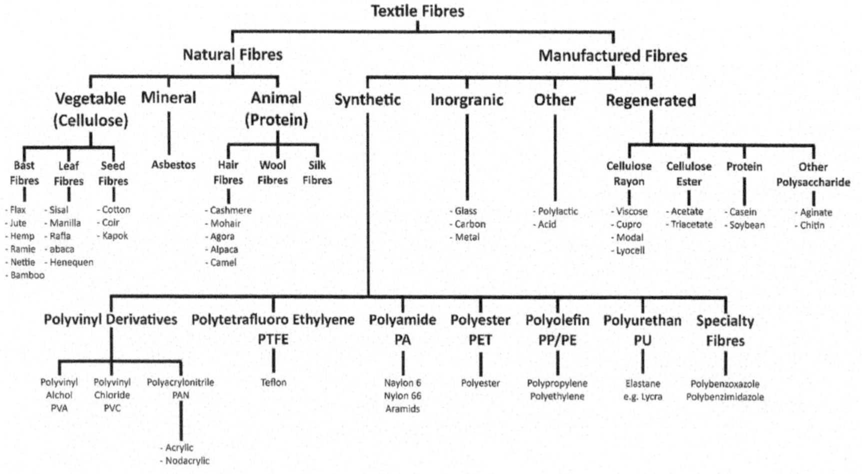

Fig. 1 Classification of textile fibres [7]

land for cultivation, it also requires labour, water, energy, pesticides, artificial fertilizers etc. Synthetic fibre processing requires energy and its embodied energy is more than natural fibre embodied energy. However natural fibre production needs a huge amount of land. Therefore, the Ecological Footprint of natural fibre production is more than synthetic fibre production. Kiran et al. examine using biopolymers in the textile industry offers financial and environmental benefits. These polymers, derived from natural sources like plants and microbes, are compostable, biodegradable, and recyclable, making them environmentally friendly alternatives to traditional synthetic materials. Biopolymers can reduce plastic waste in textiles and replace petroleum-based synthetics, which take centuries to break down [7].

The details of different types of fibres used in the textile industry are as follows:

(a) **Natural Fibres**

Natural fibre consists of plant fibres and animal proteins. These types of fibres generally come from primary or secondary meristematic tissue [8]. The different types of natural fibres used for textile fabric production are mentioned as follows:

(i) **Cotton**

Cotton fibres are also known as breathable fibres and absorbent. It can blend with other natural or manufactured fibres (like nylon, linen, wool, and polyester) and produce good-quality fabric. According to the Cotton Association of India, cotton fibre production is expected to reach around 7 million tons by 2030 because of the growing demand for cotton fabric [9].

(ii) **Wool**

Wool fibres (protein) are obtained from sheep skin and it used for textile fabric manufacturing. Its unique structure properties are used for making winter clothes [10]. The annual wool fibre production in India is around 1.2 million kg. Various studies reported that wool textiles have almost threefold higher embodied emissions than cotton fabrics. The reported embodied emissions of pure wool fabrics and blended wool-polyester fabrics are 14.07 kg-CO_2e/kg and 13.55 kg-CO_2e/kg; respectively [11].

(iii) **Silk**

Silk fibres are eco-friendly, biodegradable and highly crystalline with a well-aligned structure. Silk fibres are extracted from silkworm cocoons. It has higher tensile strength than other man-made fibres as well as good thermal stability (up to 140 °C). The densities of silk fibres are in the range of 1320–1400 kg/m^3. Indian silk production contributes 18% of the total global silk fibre production [12].

(iv) **Jute**

Jute serves as a biodegradable and sustainable textile fibre that holds great promise for environmentally conscious textile production. In India, jute fibre is the second most produced natural fibre, with an impressive output of 3.6 million tons, just behind cotton. Its lightweight quality, combined with its strong mechanical resistance, allows jute to grow naturally to significant sizes, making it an ideal option for structural applications [13].

(b) **Manufactured Fibres**

A significant portion of textile applications rely on synthetic polymers as their raw materials. However, these polymers are nonbiodegradable and are sourced from non-renewable resources. The unique structure of these synthetic polymers, known for their stability and strength, allows for their versatile use. However, this same structure also poses a challenge as it makes these polymers resistant to degradation, leading to their persistence in the environment. Furthermore, the synthesis of certain polymeric materials involves the use of toxic compounds or results in the production of toxic by-products.

(i) **Nylon**

Aliphatic polyamide is commonly known as nylon and is distinguished by the number of aliphatic carbons it possesses. Despite this, nylon is typically referred to as polyamide fibre. The most prevalent nylons used in fabric production are Nylon 66 and Nylon 6. [14].

(ii) **Polyester**

Polyethylene terephthalate (PET), the highly effective synthetic fibre, consists of ester links connecting aliphatic (ethylene diol) and aromatic (terephthalic acid) groups.

(iii) **Polyacrylonitrile**

Due to its atactic nature, polyacrylonitrile experiences the random protrusion of large cyanic groups from its main chain, impeding the achievement of a dense packing structure. As a result, the crystallinity of polyacrylonitrile fibre is relatively low.

(iv) **Polyolefin**

Polyethylene and polypropylene are two prominent polyolefin fibres that are accessible in the market. Polyethylene is distinguished by a basic chemical composition comprising exclusively of methyl groups.

1.1 Textile Production Stages

Different types of production stages are involved in fabric production. The main substages are the fibre process (raw fibre production and cleaning), spinning process (yarn production), weaving process (generally prepared in a power loom in the city), knitting process, nonwoven process, wet process, and clothing manufacturing. The general production process of textile products is given in Fig. 2.

1.2 Indian Textile Industry

The Indian textile industry has a wide range from hand-spun khadi to a highly modernised mill sector. The strength of the Indian textile industry is its production base of a wide range of fibre/yarns. The detailed classification of the textile industry

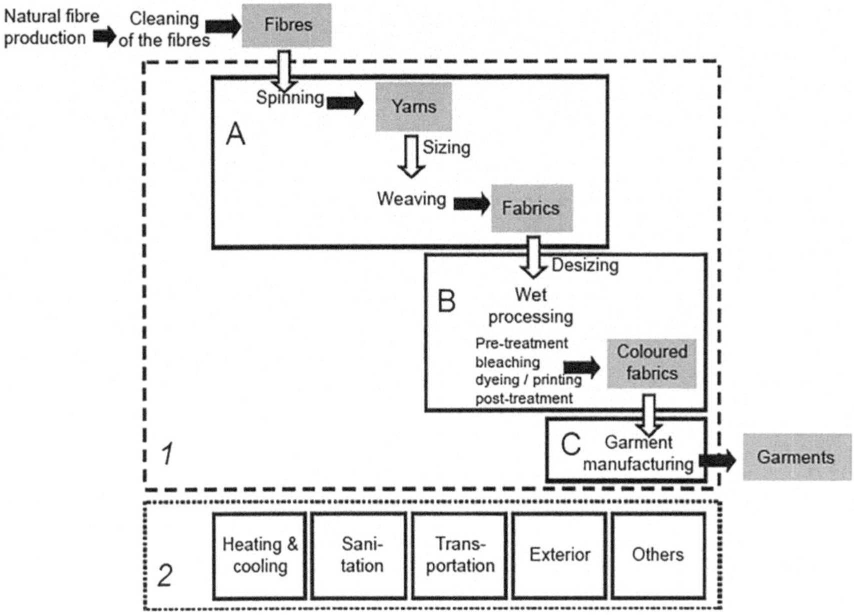

Fig. 2 General production processes of textile products

in India is shown in Fig. 3. Indian textile industry can be classified into two groups based on fibres used in textile production: (1) natural fibres like cotton, wool, silk jute etc. and (2) man-made fibres like rayon, synthetics blended fibres etc.

1.3 Environmental Impact of Fabrics

The environmental impact of fabric production is significant, the cradle-to-gate carbon footprints of fabrics range from about 0.7–36 $kgCO_{2e}$ per kg of fabric production [1]. Figure 4 shows the cradle-to-gate emission factor (per kg) of different textile processes [15]. Reducing resource and energy consumption in fabric production may move towards more industrial sustainability and reduce the country's textile industry's dependence on resources and energy imports.

The objective of this study is to apply the Ecological Footprint (EF) assessment methodology to assess the textile industry (mostly based on handloom) of Malegaon city (Nashik), India. The Ecological Footprint assessment based on the cradle-to-gate approach of fabric production in the Malegaon textile industry has been considered in the study. It also includes the identification of mitigation options to increase energy efficiency, promoting at the same time the reduction of CO_2 emissions and simultaneously decreasing the Ecological Footprint of the textile industry.

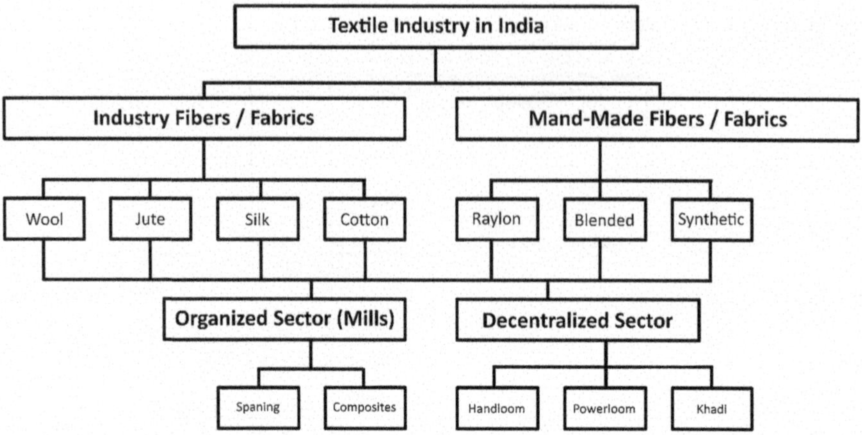

Fig. 3 Indian textile fabric production (self-made)

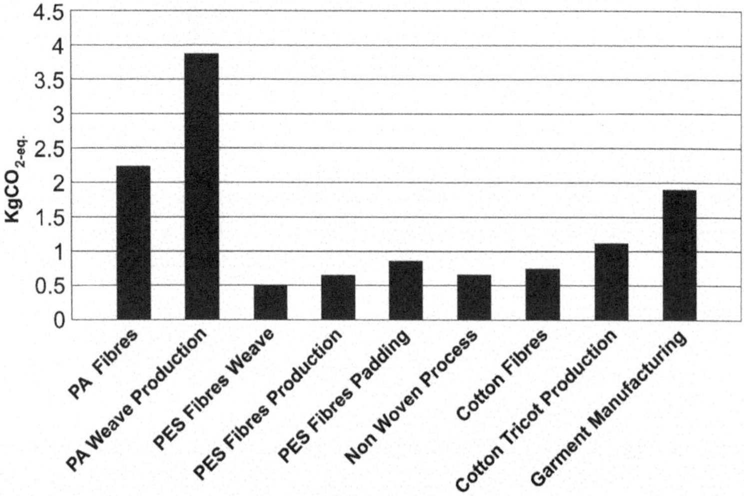

Fig. 4 The emissions factor (per kg) of different textile processes (self-made)

2 Materials and Methods

The study is focused on the ecological footprint (EF) assessment of textile processes (cotton fabrics and synthetic fabrics using handloom) (Fig. 5). Malegaon City is home to approximately 0.125 million plain power looms distributed across 6000 textile manufacturing units. The textile industry in Malegaon focuses on producing 30–35% cotton grey fabrics like Cambric and Poplin, which serve both industrial and clothing purposes. Moreover, 60% of the fabrics manufactured are based on man-made fibres. The grey fabrics produced in Malegaon's textile industry are considered semi-finished and require further processing [16].

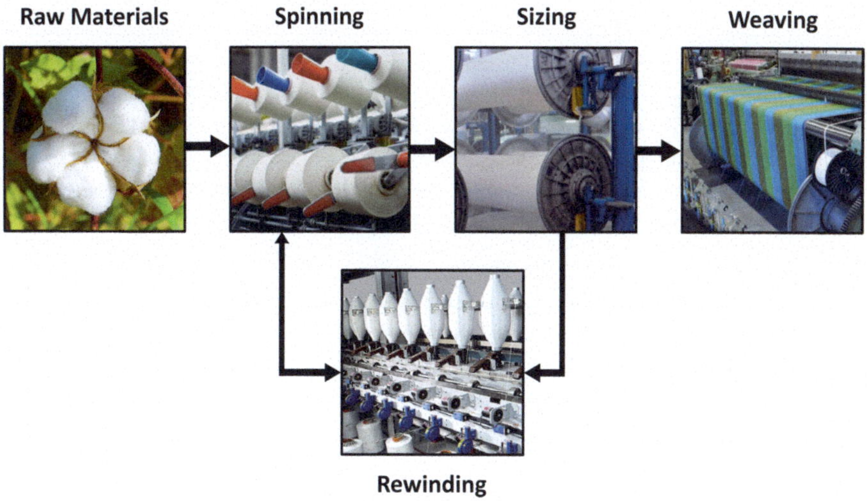

Fig. 5 The flow of materials and energy for natural (cotton) fabric production

3 Ecological Footprint of the Malegaon Textile Industry

The study is focused on the ecological footprint (EF) indicator as a means to evaluate the sustainability and environmental performance of Malegaon city textile processes (cotton fabrics and synthetic fabrics using handloom). The main concepts involved in ecological footprint are described (e.g., embodied energy, energy productivity, natural productivity, equivalence factors for land type, etc.). The Ecological Footprint of cotton fabric and synthetic fabric production are 16.64–18.17 gha/ton and 10.2–11.87 gha/ton; respectively.

3.1 Ecological Footprint of Raw Materials Production

Every raw material has some amount of environmental impact. In this study, the Ecological Footprint of raw materials production has been estimated based on two parameters: (1) the natural ecological imprint of textile raw materials, and (2) the manufacturing ecological imprint of textile raw materials. The environmental impact of textile raw materials production has been calculated by Eq. 1 [17]:

$$\text{EF of textile materials production} = \underbrace{\sum\left(\frac{Cfi}{Yfi}\right).e_i}_{(\textit{Natural ecological imprint})} + \underbrace{\sum\left(\frac{(Cfi + Ri).Emi}{Af/(1 - Aoc)}\right).e_{CO2land}}_{(\textit{Manufacturing ecological imprint})} \quad (1)$$

Where,

C_{fi} is natural fibre consumption in textile fabric production (such as cotton, silk, jute, & wool etc. Y_{fi} is the productivity of respective natural fibre. e_i represents the equivalence factor of different productive lands. E_{mi} is the embodied emission of fibre production. A_f absorption factor of forests is considered to be 2.7 tCO_2/ha [18]. A_{oc} is the fraction of annual oceanic emission sequestration (i.e. 0.3) [19]. The e_i is the equivalence factor of bio-productive lands [20]:

- Cropland ($e_{cropland}$) = 2.52 gha/ha,
- Carbon absorption/Energy land ($e_{CO2\ land}$) = 1.28 gha/ha,
- Pasture land ($e_{pasture\ land}$) = 0.43 gha/ha,
- Forest land (e_{forest} land) = 1.28 gha/ha,
- Marine/Sea-productive land ($e_{marine\ land}$) = 0.35 gha/ha

The details of different types of fibres' embodied energy/emissions and productivity are mentioned in Table 1. The Ecological Footprint of different fibres are estimated in this study and the values are also mentioned in Table 1.

The Ecological Footprint of cotton fibre production is estimated in the range of 5.71–6.11 gha/ton. The Ecological Footprint of synthetic (polyester) fibre production is estimated as 2.43 gha/ton. It shows that the environmental impact of polyester fibre production is less than 50% of the cotton fibre production.

Table 1 Embodied Energy and Ecological Footprint of the different textile raw materials

Types of fibres	Embodied energy per kg fibre (MJ/kg) [20]	Productivity	Land type	Ecological footprint of fibre (gha/ton)
Cotton khadi	34.5	0.5 tons of cotton/ha [21]	Cropland & CO_2 land	5.71
Cotton mill woven	44	0.5 tons of cotton/ha [21]	Cropland & CO_2 land	5.89
Cotton conventional	55	0.5 tons of cotton/ha [21]	Cropland & CO_2 land	6.11
Flax	10	1.2–1.5 tons/ha [22]	Cropland & CO_2 land	2.29
Wool	63	22–25 kg/ha [23]	Pasture land & CO_2 land	20.77
Viscose	100	–	CO_2 land	1.94
Polypropylene	115	–	CO_2 land	2.23
Polyester	125	–	CO_2 land	2.43
Acrylic	175	–	CO_2 land	3.40
Nylon	250	–	CO_2 land	4.85

3.2 *Ecological Footprint of Textile Processes*

(i) **Yarn Production**

Yarn production consists of a series of steps that are essential for converting fibres into yarn. Spinning is a crucial part of yarn manufacturing, where fibres are transformed into yarns. Specialized machines are employed at different stages of yarn manufacturing to achieve superior quality in yarn production. The emissions related to cotton yarn spinning are 3.84–4.05 kgCO_{2e}/kg of cotton fibre and for polyester yarn spinning are 2.8–4.56 kgCO_{2e}/kg of fibre [21, 24]. The Ecological Footprint of cotton (fibre) yarn production in spinning mills is in the range of 1.27–1.34 gha per ton of yarn. However, the Ecological Footprint of polyester (fibre) yarn production is in the range of 0.93–1.51 gha per ton of yarn.

(ii) **Sizing Process**

Following the warping process, sizing is carried out as a crucial preparatory step in weaving. This process holds immense importance and is considered critical. Sizing involves the application of a thin film of adhesive and binder onto the yarn surface, which effectively enhances the weavability of the yarn. Consequently, the strength of the warp yarn is notably increased. The power loom's beams are prepared in the sizing process. In the Malegaon textile industry, 600–700 kg of wood (Ecological Footprint of wood consumption (as a fuel) is 6.5 gha/ton [26]) is consumed (in the boilers) for one ton of beams during the sizing process and 10 labourers are required. The image of the sizing industry located in Malegaon is shown in Fig. 6.

Fig. 6 The image of the sizing industry located in Malegaon (self-made)

This section considered only energy consumption impact to estimate the sizing process of the textile industry. The impact of labour is incorporated in Sect. 3.4. On that basis, the Ecological Footprint of beam preparation is estimated as 3.9–4.56 gha per ton of yarn (beam production) sizing operation.

(iii) **Weaving Process (based on Power loom)**

The warp yarn and weft yarn interlock with each other known as the weave (Fig. 7). The weaving process consists of interlacing two sets of yarns at right angles to each other in a power loom. The yarns that are used along the fabric length are called warp yarns and the yarns that are used across the fabric length are called weft yarns. The details of yarn thread and energy consumption for fabric production through the power loom are mentioned in Table 2.

Fig. 7 The image of the weaving process through a power loom (self-made)

Table 2 The details of the weaving process (power loom)

Fibre type	Specification	Time (h/ton)	Energy (kWh/ton)
Polyester	Weft = 0.21 mm Warp = 0.21 mm Fabric width = 46 inch Weight/length = 57–59 gm/m	3508–3390	1264.5–1308.5
Cotton	Weft = 0.21 mm Warp = 0.21 mm Fabric width = 39 inch Weight/length = 88–100 gm/m	2272–2416	847.7–901.4

The Ecological Footprint of Indian grid electricity generation is reported as 2.72×10^{-4} gha/kWh (ignoring factors other than emissions) [26]. The Ecological Footprint of energy consumption for yarn weaving in a power loom is 0.34–0.36 gha per cotton fabric production. The Ecological Footprint of energy consumption for yarn weaving in a power loom is 0.23–0.25 gha per polyester (synthetic) fabric production.

3.3 Ecological Footprint of Transportation Impact

The transportation impact of the textile industry depends on raw fibre transportation from sources (such as farming land of cotton) to spinning plants and spinning to sizing plants and seizing to weaving (power loom). The Ecological Footprint of Transportation Impact is estimated by Eq. 2

$$\text{Ecological Footprint of Transportation Impact} = \sum\left\{\left(\frac{E_{mi}.D_{mi}}{E_{HDV}}\right)f_e.\alpha_i\right\}\frac{(1-A_{oc})}{A_f} \tag{2}$$

E_{mi} and D_{mi} are represented as materials consumption (ton) and the average distance of transportation of materials (km), respectively. E_{HDV} represent the capacity of heavy-duty trucks (2.5–3.5 tons). α_i is the emission factor of fuel (3.17 kgCO_2/kg of diesel [27]), f_e (fuel consumption factor) is the average fuel efficiency of HDV (0.222 kg of fuel/km).

The Ecological Footprint of transportation is reported as $2.12–7.21 \times 10^{-5}$ gha/ton-km (for 3.5–11 ton truck capacity) [17, 28]. From cotton fibre production in different regions of Maharashtra to weaving at Malegaon city, materials transportation is considered around 100–200 km. The Ecological Footprint of transportation impact is estimated as 2.12×10^{-3} gha/ton to 14.2×10^{-3} gha/ton. However, the Polyester yarn comes to Malegaon city from Mumbai city (approx. 300 km), and the transportation impact of polyester is $6.2–21.6 \times 10^{-3}$ gha/ton of polyester fabric.

3.4 Ecological Footprint of Manpower Impact

The Ecological Footprint of labour/manpower is associated with the metabolic rate [29]. The details of the Ecological Footprint of different activities are mentioned in Table 3.

The Ecological Footprint of labour/manpower is estimated as Eq. 3 [25]:

$$\text{Ecological Footprint of labour/manpower} = \left(\frac{EF_f}{365 \times \text{daily calories burn}}\right) \times \sum (M_r.h_w)_i \tag{3}$$

Table 3 The ecological footprint of manpower activities

Activity	Metabolic rate (kcal/h) [29]	Average ecological footprint (gha/h) [25]
Machinery driver	132.0–170.8	0.88×10^{-4}
Low metabolic rate	110.3–201.9	0.91×10^{-4}
Moderate metabolic rate	203.4–310.6	1.50×10^{-4}
High metabolic rate	312.2–403.8	2.09×10^{-4}

In India, labour daily calories burned (approx.) is 2576 kcal/capita/day [30]

Table 4 The details of manpower/labour requirements in the textile industry

S. No	Operation	Types of labour	Quality	Ecological footprint of manpower per ton of fabric production
1	Farming	Unskilled	1-ton cotton production required 5 Full-Time Equivalence labour work	2.7 gha/ton
2	Spinning	Skilled	0.9 pounds (approx. 0.408 kg) of yarn per manpower working hour [33]	0.3 gha/ton
3	Sizing	Skilled	19.2–28.8 kg yarn per manpower working-hour	$5.2–7.8 \times 10^{-3}$ gha/ton
4	Weaving (power loom)	Skilled	0.052–0.061 kg of cloth production per manpower working hour	2.4–2.8gha/ton

where, the EF_f is the annual Ecological Footprint of food consumption that is reported as 0.549 gha/year/person for Indian labours [25, 31, 32]. M_r represents the metabolic rate of human activities (kcal/h). h_w represents the working hours.

Labour required in the textile industry is mentioned in Table 4. The total Ecological Footprint of manpower/labour is estimated as 5.4–5.8 gha per ton of cotton fabric production. The total Ecological Footprint of manpower/labour is estimated as 2.7–3.1 gha per ton of polyester (synthetic) fabric production.

The ecological impact of natural and synthetic fabric production in India is significant, the cradle-to-gate ecological footprints of cotton fabrics range from about 16.64 to 18.17 gha/ton of fabric production. The Ecological Footprint of cotton fibre production is more (two-fold) than polyester fibre production. This is just because cotton cultivation needs huge cropland (bioproductive land). This study reported that polyester fabric is more sustainable (in terms of ecological footprint) than cotton fabric. Reducing resources (such as bioproductive land and manpower) and fossil-based energy consumption in fabric production may move towards more industrial sustainability and reduce the country's textile industry's dependence on resources and energy imports.

4 Conclusions

The study focuses on the Ecological Footprint of the Malegaon textile industry. It can be used as a tool to analyse the environmental impact of textile processes (such as raw fibre production, spinning, sizing, and weaving). The Ecological Footprint of cotton fabric and synthetic (polyester) fabric production are 16.64–18.17 gha/ton and 10.2–11.87 gha/ton; respectively. The Ecological Footprint of cotton fibre production is more than polyester fibre production. This is because cotton cultivation needs huge cropland (bioproductive land).

The environmental impact of natural/man-made fabric production is significantly high and it can be reduced by simply reducing the demand for textile products. The study is eye-opening in the field of the textile industry and adopting simple steps (such as reducing, recycling and reuse of textile products) may help to achieve the United Nations Sustainable Development (Goal No 12—Responsible Consumption and Production).

References

1. Muthu, S. S. (Ed.). (2015). *Handbook of life cycle assessment (LCA) of textiles and clothing*. Woodhead publishing.
2. European Parliament. (2022). *The impact of textile production and waste on the environment* (infographic). Retrieved December 5, 2022. https://www.europarl.europa.eu/topics/en/article/20201208STO93327
3. texfash, "2030 Carbon footprint from fashion consumption to be highest in Australia Among G20 Nations; Lowest in India" https://texfash.com/update/2030-carbon-footprint-from-fashion-consumption-to-be-highest-in-australia-among-g20-nations.
4. Fidan, F. Ş., Aydoğan, E. K., & Uzal, N. (2023). Recent progress on life cycle sustainability assessment in textile industry: Applications for environmental, economic, and social impacts of cotton and Its derivatives. In S. S. Muthu (Ed.), *Progress on life cycle assessment in textiles and clothing. Textile science and clothing technology*. Springer. https://doi.org/10.1007/978-981-19-9634-4_7
5. India Brand Equity Foundation (IBEF), "Textile Industry Report. Dec, 2023". https://www.ibef.org/industry/textiles
6. Handbook of Life Cycle Assessment (LCA) of Textiles and Clothing. https://doi.org/10.1016/B978-0-08-100169-1.00002-2
7. Kiran, S., Rehman, F. U., Javed, S., Abrar, S., Mukhtar, N., Moied, A., et al. (2024). Ecological effects of biopolymers and their advantages for textile industry. In *Biopolymers in the textile industry: Opportunities and limitations* (pp. 241–272). Springer Nature Singapore.
8. International Cotton Advisory Committee. (2013, June 9–11). *Overview of world fiber production*. Paper presented at the 1st international conference on natural fibres Guimarães, Portugal. http://dnfi.org/wp-content/uploads/2013/08/Portugal-Paper.pdf
9. Cotton Association of India (CAI), "Monthly Report January 2024" http://www.caionline.in/
10. Kadam, V., Saini, H., Verma, K., Dubey, I., & Verma, P. (2024). Prospects of wool and woolen products. In *The Wool Handbook* (pp. 593–610). Woodhead Publishing.

11. Tekin, P., Alıcı, H., & Demirdelen, T. (2024). A life cycle analysis of a polyester–wool blended fabric and associated carbon emissions in the textile industry. *Energies, 17*(2), 312.
12. Borah, A., Borah, M. B., & Sharma, N. *Relationship between Climate and Muga Silk Production.*
13. Souza, A. T., Pereira, F., Neuba, L. D. M., Clay, A., Rangel, A., & De Azevedo, G. (2020). Caranan fiber from Mauritiella Armata Palm Tree as novel reinforcement for epoxy composites. *Polymers, 12*, 2037.
14. Kajiwara, K., & Ohta, Y. (2009). Synthetic textile fibers: Structure, characteristics and identification. In *Identification of textile fibers* (pp. 68–87). Woodhead Publishing.
15. Roos, S., Sandin, G., Zamani, B., & Peters, G. M. (2015). *Environmental assessment of Swedish fashion consumption. Five garments – Sustainable futures. A Mistra Future Fashion Project 2 Report.* Chalmers University of Technology.
16. www.malegaononline.in/city-guide/major-industries-in-malegaon
17. Husain, D., & Prakash, R. (2019). Ecological footprint reduction of built envelope in India. *Journal of Building Engineering, 21*, 278–286.
18. Mancini, M. S., Galli, A., Niccolucci, V., Lin, D., Bastianoni, S., Wackernagel, M., & Marchettini, N. (2016). Ecological footprint: Refining the carbon footprint calculation. *Ecological Indicators, 61*(2), 390–403.
19. Scripps Institution of Oceanography (SIO), "The Keeling Curve" (2017) https://scripps.ucsd.edu/programs/keelingcurve/2013/07/03/how-much-co2-can-the-oceans-take-up/. Accessed on Apr 2023.
20. Global Footprint Network, (GFN). (2016). http://data.footprintnetwork.org/analyzeTrends.html?cn=100&type=EFCtot. Accessed on Nov 2023.
21. Blaise, D., & Kranthi, K. R. (2019). Cotton production in India. *Cotton Production*, 193–215.
22. Discover Natural Fibres Initiative., https://dnfi.org/flax-fibres. Accessed 08 May 2024.
23. Devaux, C. A. (2019). *Wool production, systematic review of life cycle assessment studies.*
24. Steinberger, J. K., Friot, D., Jilliet, O., & Erkman, S. (2009). A spatially explicit life cycle inventory of the global textile chain. *The International Journal of Life Cycle Assessment, 14*, 443–445. https://doi.org/10.10007/s11367-009-0078-48
25. Ansari, Y., Husain, D., Haadi, S. M., Haloi, J., & Prakash, R. (2023). Life cycle ecological footprint of building: A case study of low-rise tropical residential building. *International Journal of Environmental Science and Technology, 20*(9), 9779–9794.
26. Biswas, A., Husain, D., & Prakash, R. (2021). Life-cycle ecological footprint assessment of grid-connected rooftop solar PV system. *International Journal of Sustainable Engineering, 14*(3), 529–538. https://doi.org/10.1080/19397038.2020.1783719
27. European Environment Agency (EEA). (2013). *EMEP/EEA air pollutant emission inventory guidebook – 2013.* http://www.eea.europa.eu/publications/emep-eea-guidebook-2013
28. Nawandar, V., Husain, D., & Prakash, R. (2021). Ecological footprint assessment and its reduction for packaging industry. In S. S. Muthu (Ed.), *Assessment of ecological footprints. Environmental footprints and eco-design of products and processes.* Springer. https://doi.org/10.1007/978-981-16-0096-8_2
29. NTP 1011. (2014). Determinacion del metabolismo energetico *mediante tablas*/Determination of energetic metabolism through tables.
30. Statistic division of the FAO, FAOSTAT. (2021). https://www.fao.org/faostat/en/#data/FS. Accessed 30 Jan 2023.

31. Husain, D., & Prakash, R. (2019). Life cycle ecological footprint assessment of an academic building. *Journal of The Institution of Engineers (India): Series A, 100*, 97–110.
32. Husain, D., Prakash, R., & Ahmad, A. (2022). Life cycle ecological footprint reduction for a tropical building. *Advances in Civil Engineering, 2022*.
33. Wolcott, S., & Clark, G. (1999). Why nations fail: Managerial decisions and performance in Indian cotton textiles, 1890–1938. *The Journal of Economic History, 59*(2), 397–423.

Analyzing the Ecological Footprint of Dry Chemical Silage Derived from Red Tilapia (Oreochromis spp.) Viscera – A Case Study Utilizing Ecological Footprint Methodology

Yhoan Sebastian Gaviria Gaviria, Jairo Andrés Camaño Echavarría, and José Edgar Zapata Montoya

1 Roadmap of the Chapter

In Fig. 1, the roadmap of the chapter is shown, identifying each of the stages and the main parts of these stages. The chapter begins with an introduction about fish waste increasing and the environmental issue generated. Following with silage as alternative source of protein and drying as conservation method of this. Subsequently, it discusses the methodology for quantifying environmental impact through ecological footprint analysis. The next section is the methodology of the study, starting with the system delineation, the mathematical models for environmental quantification, and the characterization of the dried silage obtained at the end of the process. Next, the results of the study are presented, comparing and discussing them with other similar studies, to finally conclude.

Y. S. Gaviria Gaviria · J. E. Zapata Montoya (✉)
Nutrition and Food Technology Research Group, NUTEC, Universidad de Antioquia, Medellín, Colombia
e-mail: Yhoan.gaviria@udea.edu.co; edgar.zapata@udea.edu.co

J. A. Camaño Echavarría
Laboratory Reactions and Chemical Engineering, Université de Lorraine, CNRS, LRGP, Nancy, France
e-mail: jairo-andres.camano-echavarria@univ-lorraine.fr

S. S. Muthu (ed.), *Ecological Footprint of Industrial Spaces and Processes*, Environmental Footprints and Eco-design of Products and Processes,
https://doi.org/10.1007/978-3-031-69047-1_3

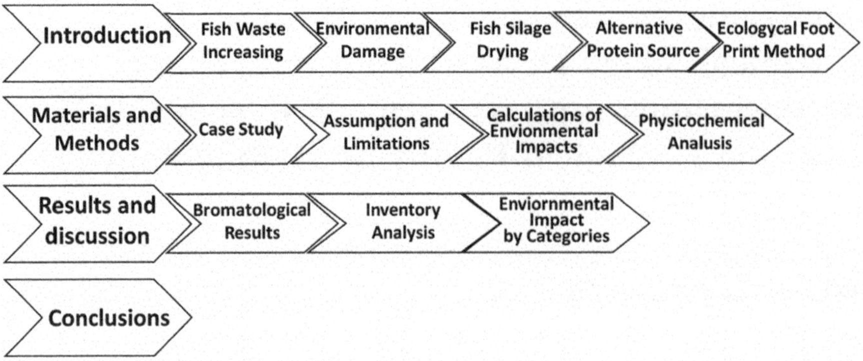

Fig. 1 Roadmap of the chapter

2 Introduction

The production of fish worldwide has increased over the years, due to an increase in the population and in consumer interest for this type of product as a result of its nutritional benefits [19, 35]. This is how aquaculture has recently become one of the industries with the highest development indices [10], reaching 171 million tons in 2018 according to the United Nations Food and Agriculture Organization [8]. In Colombia, it is estimated that fish production increased by 9% in 2018, with red tilapia (*Oreochromis* spp.) representing 62% of total production [32]. This increase in production is accompanied by the generation of waste, which represents 65% of total production, causing environmental deterioration due to inadequate final disposal [31].

In this context, silage represents an alternative use for some of the fish farming industry's most polluting waste [42]. Thus, silage has been used to process trout waste [15], red tilapia [7], and other species. This process is carried out by adding lactic acid microorganisms, organic or inorganic acid, which reduce the level of microorganisms that cause deterioration and activate endogenous enzymes, causing protein lysis, favoring the formation of peptides and amino acids of interest [33]. This product is generally obtained in liquid form, however, there are disadvantages to using it in dry diets due to its moisture content [28]. Therefore, drying processes are often used, increasing the concentration of its macro components such as protein, fat, and some organic acids [33]. Although silage is an alternative to using by-products, it comprises a series of stages that involve different factors that directly or indirectly cause environmental deterioration [38], especially when the aim is to create a dry product [7].

The economic growth of many nations at the global level is associated with the depletion of natural resources, with the current human demand for water, food, infrastructure, and energy exerting ecological pressure on the planet. This leads to a variety of factors that damage the environment, such as the depletion of natural resources, the generation of waste and the emission of greenhouse gases

[2]. Activities that involve the use of these resources gradually increase to the level that they cause environmental damage; this is how the availability of natural resources (biocapacity) is limited, as there is insufficient capacity to counteract the impact caused by the creation of waste [36].

As a result, the world has faced great challenges to the preservation of natural resources over the past few decades [20]. Currently, issues concerning sustainable development are becoming more prevalent and interest in these is increasing; this is reflected in the energy and environmental policy changes that have been occurring in many countries. [36]. As a result of this, a variety of methods have emerged that allow environmental degradation to be monitored, evaluating the impact of various processes. This includes monitoring ecological footprints, which is a method capable of generating a clear perspective of processes' and even countries' sustainability [39].

Ecological footprint is a method of calculating the effect of anthropogenic activities on land, water, forest products, infrastructure and carbon footprints [1]. This method is currently one of the most widely used sustainability indicators because it produces comparable, reliable and comprehensive responses [2, 25]. It also estimates the effects of human activity on the environment, establishing the amount of hectares of land and water worldwide that are necessary to produce and mitigate the waste generated in the consumption of natural resources [2]. The ecological footprint is therefore used to evaluate the environmental impact in optimization processes, allowing each stage to be analyzed, identifying the phenomena that generate the greatest impact and implementing reduction strategies that improve the process ecological impact [17, 18].

Chemical fish silage has been considered as a great alternative for the total or partial replacement of conventional protein sources in animal feed production and has been widely reported in many research studies. Due that, this process, not only offer environmental benefits due to waste reduction, but also because the silage is a protein substrate with great benefits for the nutrition of different farm animals such as trout [16], broiler chickens [11, 14] and laying hens [12], among other animal species [15].

However, no studies have yet been reported that quantify the environmental impact of obtaining either liquid or dry chemical silage. Thus, the objective of this study was to determine the environmental impact created when obtaining dry chemical silage (DCS) from the viscera of red tilapia (*Oreochromis* spp.), using the ecological footprint methodology as an indicator of sustainability.

3 Materials and Methods

3.1 Case Study

The study system aims to obtain DCS from the viscera of red tilapia (*Oreochromis* spp.), which involves different processes. Fresh viscera were supplied by El Gaitero

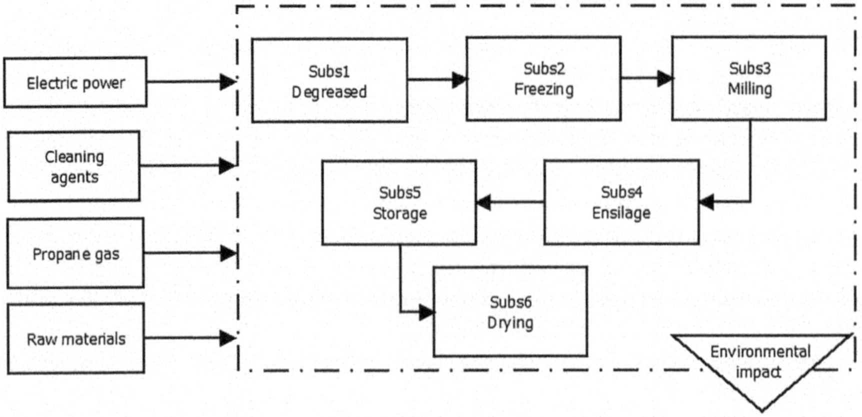

Fig. 2 Limit of the dry chemical silage manufacturing process

fish farm, located in the municipality of San Jerónimo, Antioquia-Colombia (6°26′ 30″N 75°43′40″W), and they were processed in laboratory of group Nutrición y Tecnología de Alimentos (NUTEC) located in Medellín, Antioquia-Colombia. Figure 2 shows the different stages of the process of obtaining this product, including the overall inputs and outputs throughout the system as well as its delimitation. 6 subsystems were considered which represent the main stages of the process described according to the methodology described by Arias et al. [4] and Suarez et al. [38].

Subsystem 1 (Subs1): This refers to the viscera degreasing stage, which consisted of heating the fresh viscera to 67 °C for a period of 30 min, successfully separating most of the lipid content by decantation.

Subsystem 2 (Subs2): This involves freezing the previously degreased material in order to solidify the remaining lipid content over a period of 24 h, eventually separating the frozen lipid phase from the protein-rich aqueous phase.

Subsystem 3 (Subs3): This consisted of crushing the protein-rich material obtained in subsystem 2. A knife mill (Black and Decker, USA) was used for this and a completely homogeneous material was obtained and arranged in plastic containers.

Subsystem 4 (Subs4): At this stage, the necessary additives were mixed with the homogenized viscera to initiate the hydrolysis and acidification processes. 97% sulfuric acid (Merck, Germany), 85% formic acid (Merck, Germany) and other additives were used for this, including Butyl hydroxy toluene (BHT) (Tecnas SA, Colombia) and potassium sorbate (Tecnas SA, Colombia) as an antioxidant and preservative respectively. Table 1 shows the formulation used for the chemical silage process.

Subsystem 5 (Subs5): This is the storage stage, which lasts for a period of 8 days, during which the pH decreases to values of 4 or lower. At this stage, airtight

Table 1 Chemical silage formulation

Raw material	%
Homogenized viscera	98,68
97% Sulfuric acid	0,03
85% Formic acid	1,16
BHT	0,002
Potassium sorbate	0,12

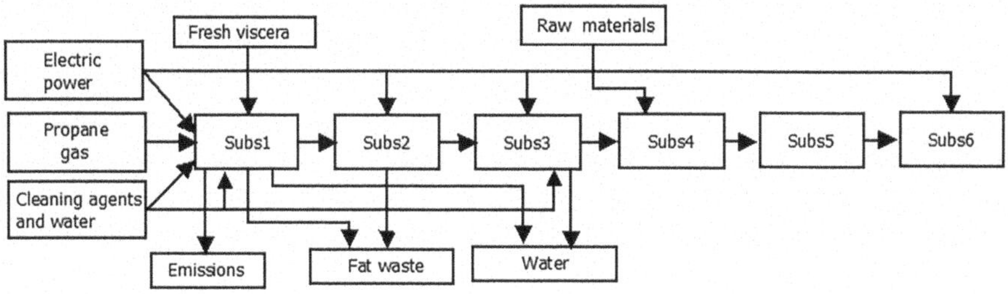

Fig. 3 Flow diagram of input and output of the dry chemical silage manufacturing process

conditions must be maintained to prevent oxygen from entering the containers and putrefactive microorganisms from spreading [38].

Subsystem 6 (Subs6): At this stage, the convective drying process was carried out in a forced air oven (Thermo Scientific™, USA) at 60 °C for 24 h until a moisture content of 10% is reached.

Meanwhile, Fig. 3 shows all the subsystems with the inputs and outputs specified in each operation, in order to evaluate the environmental impact of each of these. These include electric power and caloric energy inputs, cleaning agents, additives and outputs such as lipid waste generation and cleaning water.

3.2 *Assumptions and Limitations*

The following premises were considered for the present study:

1. The DCS manufacturing process was carried out on El Gaitero fish farm, located in the municipality of San Jerónimo, Antioquia-Colombia, where the fresh viscera come from. This was done to avoid the negative environmental impact caused by transport, as this process is intended to be carried out on the fish farms where the fish originate.
2. The impact of buildings was not taken into account in our calculations as they have a long service life.
3. The environmental impact of industrial equipment was not included in the environmental analysis due to limitations in obtaining data from manufacturing

processes. However, it has been shown that these do not significantly affect the environmental impact because they have long service lives [34].
4. Subs6 considers water vapor molecules to be the only compounds emitted into the environment as a result of the drying process, since at this stage the system does not reach the temperatures required to evaporate other compounds.

3.3 Calculation of Environmental Impact

To analyze the environmental impact of the process of manufacturing dry chemical silage, the ecological footprint study was used. This methodology was carried out as established by Gwehenberger and Narodoslawsky, [17, 18] and Krotscheck and Narodoslawsky [27], in which calculation models are used to estimate the ecological impact of each activity shown in Fig. 2. This makes calculations more precise as a large set of information is required and limitations that exist due to the use of international databases are reduced [30]. To calculate the ecological footprint of the process, one ton of DCS was the fundamental unit. The environmental impact values of the inputs and outputs of each stage of the process were normalized to the same unit (Ha / ton) which indicates the number of hectares of forest necessary to replenish the CO_2 produced during the process.

Seven calculation models based on IPCC (Intergovernmental Panel on Climate Change) [23] reports were used. Equation 1 was used to calculate the environmental impact (EI) caused by the organic compounds (*OC*) used or generated within the process. This was carried out in accordance with the provisions of the Intergovernmental Panel on Climate Change (IPCC) guidelines, where AOR corresponds to the amount of organic residue, CFM to the correction factor for methane gas (CH_4) that depends on the process associated with solid waste management in a particular sector [23], DOC represents the fraction of degradable organic carbon in the residue that can be subject to biochemical decomposition [22], DCO_F is the fraction of DCO not assimilated or that degrades very slowly [23], F and R correspond to the fraction of CH_4 in the landfill gas and the fraction of CH_4 recovered respectively. The CH_4 oxidation factor is represented by OX, the global warming potential of methane (GWP) was used for a period of 100 years [24] and FF corresponds to the CO_2 fixing factor for the region established within the environmental analysis [21]. The coefficients 16 and 12 are the molecular weights of methane CH4 and C, respectively.

$$EI = \sum_{OC=1}^{n} AOR\left(\frac{16 * (CFM * DCO * DCO_F * (F - R) * (1 - OX) * GWP)}{12 * FF}\right) \tag{1}$$

The environmental impact of the input of electrical power at different stages was calculated using Eq. 2, where EE_G corresponds to the expenditure of electrical power

(e) at each stage of the process and EEE is the factor of CO_2 emissions caused by one kW-h of energy.

$$EI = \sum_{e=1}^{n} EE_G * \frac{EEE}{FF} \tag{2}$$

In subs1 the degreasing process is carried out through heating with propane gas. Its impact is calculated using Eq. 3, where RV_{C3H8} is the required volume of gas for the process, EB_p is the embodied energy of the process of obtaining propane gas, D_R is the distance traveled by transport, R_c is fuel efficiency and C is the amount of propane gas, while CO_2 emissions from fuel are represented by EC. In addition to this, this subsystem also involves CO_2 generation due to the propane gas combustion process, whose impact was calculated using Eq. 4, where CF_{C3H8} corresponds to the conversion factor for propane gas and EF_{C3H8} represents CO_2 emission factor.

$$EI = RV_{C_3H_8}\left(\frac{EB_P * EEE + \left(\frac{D_R}{R_C * C}\right) * EC}{FF}\right) \tag{3}$$

$$EI = \frac{RV_{C_3H_8} * CF_{C_3H_8} * EF_{C_3H_8}}{FF} \tag{4}$$

For subsystems 1 and 3 it is necessary to include an input flow corresponding to cleaning water. The impact caused by the supply of water was calculated using Eq. 5, in which W_L corresponds to the liters of water required in the process (a) and EB_W is the embodied energy for the water supply. However, the use of water is accompanied by an environmental impact associated with its eventual disposal, which is calculated using Eq. 6, where RW_L corresponds to the volume of residual water obtained in the cleaning process, CO_{BOD5} to the biochemical demand for oxygen from the degradable fraction of wastewater [23], and MPC_{CH4} to the maximum methane production capacity of this fraction.

$$EI = \sum_{a=1}^{n} W_L\left(\frac{EB_w * EEE}{FF}\right) \tag{5}$$

$$EI = \sum_{a=1}^{n} RW_L * \left(\frac{CO_{DBO5} * MPC_{CH4} * CFM_{CH4} * PCG}{FF}\right) \tag{6}$$

For the environmental impacts associated with production ingredients and other input flows in the different subsystems, Eq. 7 is used, where RQ_I is the required quantity of product (pi), EB_{PI} corresponds to the embodied energy of the manufacture of the product and C_I corresponds to the maximum amount in transporting the product.

$$EI = RQ_I\left(\frac{EB_{PI} * EEE + \left(\frac{D_R}{R_C * C_I}\right) * EC}{FF}\right) \tag{7}$$

3.4 *Physicochemical Analysis*

The analytical methods established by the AOAC (*Association of Official Analytical Chemists*) [3] were used to determine the bromatological characterization of fresh viscera and wet and dry chemical silage. The moisture content was determined using the methodology established in standard 930.15, drying the sample at 105 °C (Thermo Scientific^{TM}, USA) for 8 h. The ashes were analyzed according to standard 942.05 by calcining the samples at 550 °C (Terrigeno, Colombia) for 7 h. Fat analysis was performed according to standard 920.39 (Radlys, USA). Protein was determined by the Kjeldahl method (Velp Scientifica, Italy), standard 954.01, while carbohydrates (CHO) were determined by subtracting the other components mentioned from one hundred percent [37].

4 Results and Discussion

Table 2 shows the bromatological composition of the matrices involved in the process, as well as that of the final product, highlighting the variation of important macro components such as fat and protein throughout the stages. Through the proposed elaboration process it was possible to obtain a product with the typical physical characteristics of a wet chemical silage [4, 38]. In addition, it is evident that the DCS has significant potential as alternative raw protein material for animal feed, as it meets the required protein and fat levels, making it a product of great interest as it would allow the fish farming industry's waste to be meaningfully used [7, 13].

Table 3 outlines an inventory analysis of each of the dry chemical silage obtaining process subsystems, detailing the inputs and outputs at each of the stages with the respective quantity used to produce one ton of DCS. The reported values were calculated using Eqs. 1, 2, 3, 4, 5, 6, and 7. In addition, it includes an environmental impact value for each of the material and energy flows. The fresh viscera present in

Table 2 Bromatological composition of fresh viscera, chemical silage, and dry chemical silage (%)

Nutrient	Fresh Viscera	Chemical silage	DCS
Moisture	61,36 ± 0,29	81,27 ± 0,09	10,20 ± 0,10
Protein	4,03 ± 0,10	6,85 ± 0,04	48,89 ± 0,51
Fat	32,93 ± 0,04	9,08 ± 0,01	31,58 ± 0,13
Ash	0,67 ± 0,04	2,69 ± 0,01	8,85 ± 0,02
CHO	0,98 ± 0,09	0,11 ± 0,02	0,48 ± 0,04

Table 3 Inventory analysis and environmental impact

Subs 1		
Inputs	**Quantity**	**[a]EI (ha/ton)**
Fresh viscera (Kg)	6589,80	−10,1482920
Electric power (KW-h)	3	0,0000587
Propane (m^3)	1	0,0000710
Water (L)	2000	0,1095867
Soap (Kg)	50	0,0006728
Outputs	**Quantity**	**EI (ha/ton)**
Degreased viscera (Kg)	4806,70	0
CO_2 (Kg)	6379,80	0,7595000
Cleaning water (L)	2050	1,0250000
Organic waste (Kg)	1783,10	2,7459740
Subs 2		
Inputs	**Quantity**	**EI (ha/ton)**
Degreased viscera (Kg)	4806,66	0
Electric power (KW-h)	0,0025	4,8923E-08
Outputs	**Quantity**	**EI (ha/ton)**
Degreased viscera (Kg)	4440,96	0
Organic waste (Kg)	365,70	0,5631780
Subs 3		
Inputs	**Quantity**	**EI (ha/ton)**
Degreased viscera (Kg)	4440,96	0
Electric power (KW-h)	0,1619	3,1682E-06
Water (L)	1400	0,0767107
Soap (Kg)	20	0,0006569
Outputs	**Quantity**	**EI (ha/ton)**
Homogenized viscera (Kg)	4440,96	0
Cleaning water (L)	1420	0,710000
Subs 4		
Inputs	**Quantity**	**EI (ha/ton)**
Homogenized viscera	4440,96	0
Sulfuric acid 97%	1,35	0,0002328
Formic acid 85%	52,20	3,3806E-06
BHT	0,09	1,0631E-06
Potassium sorbate	5,40	6,3787E-05
Outputs	**Quantity**	**EI (ha/ton)**
Pre-silage (Kg)	4500	0
Subs 5		
Inputs	**Quantity**	**EI (ha/ton)**
Pre-silage (Kg)	4500	0
Outputs	**Quantity**	**EI (ha/ton)**
Silage (Kg)	4500	0
Subs 6		

(continued)

Table 3 (continued)

Inputs	Quantity	EI (ha/ton)
Silage (Kg)	4500	0
Electric power (KW-h)	67266	1,3163316
Outputs	**Quantity**	**EI (ha/ton)**
Dry silage (Ton)	1	0
TOTAL (Ha/Ton)		**-2,8402475**

[a]Red environmental impact values mean environmental deterioration, and black environmental impact values mean environmental benefit

subs 1 show a value corresponding to −10,1482920 ha/ton; this is a negative value because it corresponds to a by-product of the fish farming industry that, if unused, generates an environmental impact equivalent to this magnitude. When used for preparing DCS, this impact on the environment is avoided. Similar behaviors were reported by Malakahmad et al. [29], where the environmental impact of different processes for the disposal of organic waste was determined, finding that the CO_2 emissions avoided for anaerobic digestion were considerably higher than the total CO_2 emissions of the process, resulting in negative values.

The total environmental impact delivered a value of −2,8402475 ha/ton, which indicates that the process of manufacturing DCS allows approximately 30% of CO_2 emissions to be mitigated, compared to those generated in the environment when fresh viscera are dumped in shallow landfills (10,1482920 ha/ton). This result is highly relevant as the fish farming industry generates approximately 0,07 million tons of waste per year [6], corresponding to 0,02 million tons of CO_2 equivalent, which could be mitigated with alternative harvesting processes such as silage [40]. However, during the multiple stages of the silage process, CO_2 emissions are generated due to the inputs and outputs. It is therefore important to analyze each one in detail to identify possible process improvements from an environmental perspective.

Figure 4 shows the environmental impact of the process subsystems with each of the input and output categories, as well as the total of each one. Similar inputs and outputs were consolidated within the categories; the impact of propane includes CO_2 emissions produced during combustion and the impact of water includes emissions of input water for cleaning and the eventual disposal of this. For subs 1, we can see that organic waste represents the greatest environmental impact with 59% of the total CO_2 emissions in this stage, followed by water with 24% and finally propane with 16%. Likewise, organic waste represented the highest CO_2 emission (99,99%) for subs 2. In the case of subs 3, the water intake and water disposal represent 99,91% of total CO_2 emissions at the crushing stage. Low environmental impact was registered at the silage stage (subs 4), since low percentages of raw materials were used, except for the homogenized viscera (Table 1). In subs 5, there are no CO_2 emissions to the environment, since this is the product's hermetic storage stage, where there are no inflows or outflows. Finally, in subs 6 the electrical power necessary for drying is the only factor that affects the environment.

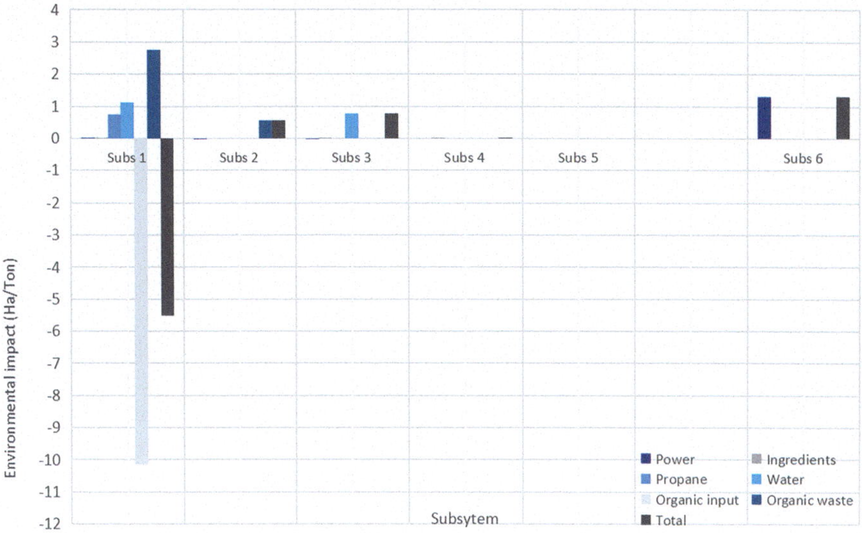

Fig. 4 Environmental impact of the subsystems of the dry chemical silage manufacturing process

Based on the foregoing, the disposal of organic waste and the energy required for the drying stage were identified as the process input and output flows that have the greatest environmental impact, with 63% and 18% respectively. This requires alternatives to be proposed to improve the process from an environmental point of view. For example, organic waste can be used to produce biogas. Bulatov et al. [5] and Kafle et al. [26] reported that the waste from the fish farming industry is a potential substrate for obtaining alternative energies such as biogas. In addition, it has been shown that lipid waste can be used in the preparation of diets for animal feed as reported by Goosen et al. [15] and Güllü et al. [16], who found that the recovered oil can be a food ingredient that does not present significant modifications in the productive variables of the evaluated species. On the other hand, conventional drying technologies are expensive and have a high environmental impact, which is why alternative technologies such as solar drying have been pursued, as these are inexpensive while not significantly modifying the characteristics of the final product [13]. Likewise, the optimal drying conditions for chemical silage made from red tilapia viscera (*Oreochromis* spp.) have been reported through the use of a direct solar dryer, achieving the physicochemical and microbiological characteristics for the product's stability [7]. As a result of the foregoing, an alternative scenario was proposed, modifying the main sources of contamination of the original process, and calculating the reduction of the environmental impact based on these modifications.

Table 4 shows a comparison between the original scenario and the alternative scenario, with the above modifications. In terms of the disposal of organic waste in the alternative scenario, a value of $-1{,}3236608$ ha/ton was obtained, which indicates a significant reduction in environmental impact compared to the original process, as the use of the waste for obtaining biogas has been used to combat adverse effects on

Table 4 Comparison of the environmental impact of the organic waste disposal and the drying process in the different scenarios

	Original scenario	EI (Ha/ton)	Alternative scenario	EI (Ha/ton)
disposal organic waste	Dumped in shallow landfills	3,3091520	Biogas production	– 1,3236608
Drying process	Convective drying	1,3163316	Solar drying	0

the environment, showing viability in sustainable processes [5]. On the other hand, at the drying stage a mitigation of 100% of CO_2 emissions caused in the original process can be achieved. This is because the solar energy used in the alternative scenario is an abundant, renewable and free source, making it one of the most promising alternatives to reduce the environmental impact of drying processes [9, 41]. Finally, the alternative scenario total environmental impact was – 8,7893919 ha/ton, corresponding to an 86% reduction in environmental impact. This was 2,8 times greater than the obtained in the original scenario (30%). This indicates that the use of waste from the fish farming industry to obtain DCS by implementing alternative technologies for its production is possible from both an environmental and technological perspective, as this waste can create added value products that could supply protein raw materials for animal feed, with the environmental aspect being one of the fundamental axes in sustainable production processes.

These results show as by adopting sustainable practices and technologies, it's possible to enhance both the environmental sustainability and economic viability of the production system. Such as the study from [11], who evaluated environmental impact from partial substitution of conventional protein for chemical silage from red tilapia viscera, in diet to broiler chickens, which results shown that was possible to reduce adverse effects, such as those caused by greenhouse gas enhancers, additionally not negatively effect on the productive variables of broiler chickens was observed. The same way to [12], who used red tilapia (Oreochromis Spp.) viscera chemical silage as an alternative protein substitute in laying hens' feed manufacturing, which result shown that strategy reduces environmental load compared to conventional waste disposal processes such as landfill disposal, making it the most environmentally favorable production system, without requiring modifications to production parameters, while also improving feed conversion efficiency.

5 Conclusions

The implementation of the ecological footprint as an environmental assessment method of a process to obtain dry chemical silage from red tilapia viscera (*Oreochromis* spp.) allowed the true environmental impact of the use of these residues to be quantified, identifying the stages and input and output flows with the highest levels of CO_2 emissions, which was useful for implementing new alternative technologies that optimize deterioration, resulting in a technologically and environmentally stable process. In addition to this, the organic waste involved in the process of obtaining DCS proved to be the main source of CO_2 emissions. However, it has been shown that this has a high potential to produce alternative energy such as biogas.

The production of DCS with the use of alternative drying energies such as solar drying, proved to be a sustainable process from an environmental perspective, in addition to being a product with added commercial value with ideal characteristics to be used as a substitute for protein raw materials in the animal feed industry.

Future research on the implementation of other environmental quantification methods in the process of obtaining chemical fish silage and their comparisons, as well as the quantification of the environmental impact in feeding different animal species with chemical silage could be of interest.

References

1. Ahmed, Z., Asghar, M. M., Malik, M. N., & Nawaz, K. (2020). Moving towards a sustainable environment: The dynamic linkage between natural resources, human capital, urbanization, economic growth, and ecological footprint in China [Forthcoming]. *Resources Policy, 67*(April), 101677. https://doi.org/10.1016/j.resourpol.2020.101677
2. Ahmed, Z., & Wang, Z. (2019). Investigating the impact of human capital on the ecological footprint in India: An empirical analysis. *Environmental Science and Pollution Research, 26*(26), 26782–26796. https://doi.org/10.1007/s11356-019-05911-7
3. AOAC. (2000). *Official methods of analysis of AOAC international* (17a editio). Association of Official Analysis Chemists International.
4. Arias, L., Gómez, L. J., & Zapata, J. E. (2017). Efecto de temperatura-tiempo sobre los lípidos extraídos de VÍsceras de tilapia roja (Oreochromis sp.) utilizando un proceso de calentamiento-congelación. *Informacion Tecnologica, 28*(5), 131–142. https://doi.org/10.4067/s0718-07642017000500014
5. Bulatov, N. K., Sarzhanov, D. K., Elubaev, S. Z., Suleymenov, T. B., Kasymzhanova, K. S., & Balabayev, O. T. (2019). Model of effective system of processing of organic wastes in biogas and environmental fuel production plant. *Food and Bioproducts Processing, 115*, 194–207. https://doi.org/10.1016/j.fbp.2019.03.005
6. Cakar, B., Aydin, S., Varank, G., & Ozcan, H. K. (2020). Assessment of environmental impact of FOOD waste in Turkey. *Journal of Cleaner Production, 244*, 118846. https://doi.org/10.1016/j.jclepro.2019.118846
7. Camaño, J. A., Rivera, A. M., & Zapata, J. E. (2020). Efecto del espesor de película y de la ubicación de la muestra en un secador solar directo, sobre la cinética de secado de ensilado de vísceras de tilapia roja (Oreochromis sp). *Información Tecnológica, 31*(1), 53–66. https://doi.org/10.4067/s0718-07642020000100053
8. FAO. (2018). The State of world fisheries and aquaculture. *Meeting the sustainable development goals*. https://doi.org/Licencia:CCBY-NC-SA3.0IGO
9. Fudholi, A., Sopian, K., Bakhtyar, B., Gabbasa, M., Othman, M. Y., & Ruslan, M. H. (2015). Review of solar drying systems with air based solar collectors in Malaysia. *Renewable and Sustainable Energy Reviews, 51*, 1191–1204. https://doi.org/10.1016/j.rser.2015.07.026

10. Fuentes-Santos, I., Cubillo, A. M., & Labarta, U. (2017). A bioeconomic approach to optimize mussel culture production. *Reviews in Aquaculture, 9*(2), 125–140. https://doi.org/10.1111/raq.12108
11. Gaviria, Y.S., Figueroa, O.A., y Zapata, J.E., Efecto de la inclusión de ensilado químico de vísceras de tilapia roja (Oreochromis spp.) en dietas para pollos de engorde sobre los parámetros productivos y sanguíneos. Información tecnológica, 32(3), 79-88 (2021), https://doi.org/10.4067/S0718-07642021000300079.
12. Gaviria, Y.S., Londoño, F.L.F., y Zapata, J.E., Effects of CS of red tilapia viscera (Oreochromis spp.) as a source of protein on the productive and hematological parameters in isa-brown laying hens (Gallus gallus domesticus), https://doi.org/10.1016/j.heliyon.2020.e05831, Heliyon, 6(12) (2020).
13. Goddard, J. S., & Perret, J. S. M. (2005). Co-drying fish silage for use in aquafeeds. *Animal Feed Science and Technology, 118*(3–4), 337–342. https://doi.org/10.1016/j.anifeedsci.2004.11.004
14. Gomez, G. M., Ortiz, M. A., Perea, C., & Lopez, F. J. (2014). Evaluación del ensilaje de vísceras de tilapia roja (Oreochromis spp) en alimentación de pollos de engorde. *Biotecnología en el sector agropecuario y agroindus-trial, 12*(1), 106–114.
15. Goosen, N. J., de Wet, L. F., Görgens, J. F., Jacobs, K., & de Bruyn, A. (2014). Fish silage oil from rainbow trout processing waste as alternative to conventional fish oil in formulated diets for Mozambique tilapia Oreochromis mossambicus. *Animal Feed Science and Technology, 188*, 74–84. https://doi.org/10.1016/j.anifeedsci.2013.10.019
16. Güllü, K., Acar, Ü., Tezel, R., & Yozukmaz, A. (2014). Replacement of fish meal with fish processing by-product silage in diets for the rainbow trout, Oncorhynchus mykiss. *Pakistan Journal of Zoology, 46*(6), 1697–1703.
17. Gwehenberger, G., & Narodoslawsky, M. (2007a). The sustainable process index SPI as an engineer's tool for environmental impact assessment of processes: The sugar industry as a case study. *WIT Transactions on Ecology and the Environment, 102*, 333–340. https://doi.org/10.2495/SDP070321
18. Gwehenberger, G., & Narodoslawsky, M. (2007b). The ecological impact of the sugar sector-Aspects of the change of a key industrial sector in Europe. *Computer Aided Chemical Engineering, 24*, 1029–1034. https://doi.org/10.1016/S1570-7946(07)80196-9
19. Heller, M. C., & Keoleian, G. A. (2015). Greenhouse gas emission estimates of U.S. dietary choices and food loss. *Journal of Industrial Ecology, 19*(3), 391–401. https://doi.org/10.1111/jiec.12174
20. Ibidhi, R., Hoekstra, A. Y., Gerbens-Leenes, P. W., & Chouchane, H. (2017). Water, land and carbon footprints of sheep and chicken meat produced in Tunisia under different farming systems. *Ecological Indicators, 77*, 304–313. https://doi.org/10.1016/j.ecolind.2017.02.022
21. IDEAM. (2011). *Estimación de las emisiones de dióxido de carbono generadas por deforestación durante el periodo 2005–2010.*
22. IPCC. (1996). *Directrices para los inventarios nacionales de gases de efecto invernadero – Desperdicios – Módulo 6* (Vol. 63).
23. IPCC. (2006). Directrices para los inventarios nacionales de gases de efecto invernadero – Elimimación de desechos sólidos – Capitulo 3. In *IPCC.*
24. IPCC. (2014). Cambio climático 2014: Informe de Síntesis. In *Contribución de los Grupos de trabajo I,II y III al Quinto Informe de Evaluación del Grupo Intergubernamental de Expertos sobre el Cambio Climático.*
25. Jóhannesson, S. E., Davíðsdóttir, B., & Heinonen, J. T. (2018). Standard ecological footprint method for small, highly specialized economies. *Ecological Economics, 146*(December 2016), 370–380. https://doi.org/10.1016/j.ecolecon.2017.11.034
26. Kafle, G. K., Kim, S. H., & Sung, K. I. (2013). Ensiling of fish industry waste for biogas production: A lab scale evaluation of biochemical methane potential (BMP) and kinetics. *Bioresource Technology, 127*, 326–336. https://doi.org/10.1016/j.biortech.2012.09.032

27. Krotscheck, C., & Narodoslawsky, M. (1996). The sustainable process index. A new dimension in ecological evaluation. *Ecological Engineering, 6*(4), 241–258. https://doi.org/10.1016/0925-8574(95)00060-7
28. Madage, S. S. K., Medis, W. U. D., & Sultanbawa, Y. (2015). Fish silage as replacement of fishmeal in Red Tilapia Feeds. *Journal of Applied Aquaculture, 27*(2), 95–106. https://doi.org/10.1080/10454438.2015.1005483
29. Malakahmad, A., Abualqumboz, M. S., Kutty, S. R. M., & Abunama, T. J. (2017). Assessment of carbon footprint emissions and environmental concerns of solid waste treatment and disposal techniques; case study of Malaysia. *Waste Management, 70*, 282–292. https://doi.org/10.1016/j.wasman.2017.08.044
30. Mamouni Limnios, E. A., Ghadouani, A., Schilizzi, S. G. M., & Mazzarol, T. (2009). Giving the consumer the choice: A methodology for Product Ecological Footprint calculation. *Ecological Economics, 68*(10), 2525–2534. https://doi.org/10.1016/j.ecolecon.2009.04.020
31. Martínez-Alvarez, O., Chamorro, S., & Brenes, A. (2015). Protein hydrolysates from animal processing by-products as a source of bioactive molecules with interest in animal feeding: A review. *Food Research International, 73*(1069), 204–212. https://doi.org/10.1016/j.foodres.2015.04.005
32. Ministerio de Agricultura. (2018). *Cadena de la acuicultura.*
33. Olsen, R. L., & Toppe, J. (2017). Fish silage hydrolysates not only a feed nutrient, but also a useful feed additive. *Trends Food Science Technology, 66*, 93–97. 10.1016/j.tifs.2017.06.003
34. Perez-Martinez, M. M., Noguerol, R., Casales, B. I., Lois, R., & Soto, B. (2018). Evaluation of environmental impact of two ready-to-eat canned meat products using Life Cycle Assessment. *Journal of Food Engineering, 237*(May), 118–127. https://doi.org/10.1016/j.jfoodeng.2018.05.031
35. Pullin, R. S. V., Froese, R., & Pauly, D. (2007). Indicators for the sustainability of aquaculture. *Ecological and Genetic Implications of Aquaculture Activities*, 53–72. https://doi.org/10.1007/978-1-4020-6148-6_3
36. Sharif, A., Baris-Tuzemen, O., Uzuner, G., Ozturk, I., & Sinha, A. (2020). Revisiting the role of renewable and non-renewable energy consumption on Turkey's ecological footprint: Evidence from Quantile ARDL approach. *Sustainable Cities and Society, 57*(August 2019), 102138. https://doi.org/10.1016/j.scs.2020.102138
37. Spanopoulos-Hernandez, M., Ponce-Palafox, J. T., Barba-Quintero, G., Ruelas-Inzunza, J. R., Tiznado-Contreras, M. R., Hernández-González, C., & Shirai, K. (2010). Production of biological silage from fish waste, the smoked yellowfin tuna (Thunnus albacares) and fillet of tilapia (Oreochromis sp), for feeding aquaculture species. *Revista Mexicana de Ingeniera Qumica, 9*(2), 167–178. http://www.scopus.com/inward/record.url?eid=2-s2.0-79960206448&partnerID=40&md5=aebc679ffc5dee73f2cd9d052af5db5c
38. Suarez, L. M., Montes, J. R., & Zapata, J. E. (2018). Optimización del Contenido de Ácidos en Ensilados de Vísceras de Tilapia Roja (Oreochromis spp.) con Análisis del Ciclo de Vida de los Alimentos Derivados. *Información Tecnológica, 29*(6), 83–94. https://doi.org/10.4067/s0718-07642018000600083
39. Syrovátka, M. (2020). On sustainability interpretations of the Ecological Footprint. *Ecological Economics, 169*(December 2019), 106543. https://doi.org/10.1016/j.ecolecon.2019.106543
40. Tian, M., Gao, J., Zheng, Z., & Yang, Z. (2012). The study on the ecological footprint of rural solid waste disposal-example in Yuhong District of Shenyang. *Procedia Environmental Sciences, 16*(0), 95–101. https://doi.org/10.1016/j.proenv.2012.10.013
41. Tiwari, A. (2016). A review on solar drying of agricultural produce. *Journal of Food Processing & Technology, 7*(9). https://doi.org/10.4172/2157-7110.1000623
42. van't Land, M., Vanderperren, E., & Raes, K. (2017). The effect of raw material combination on the nutritional composition and stability of four types of autolyzed fish silage. *Animal Feed Science and Technology, 234*(October), 284–294. https://doi.org/10.1016/j.anifeedsci.2017.10.009

Ecological Footprint and Economic Assessment of Ready-Mix Concrete Production

Saud Anjum Mahevi, A. K. Kaliluthin, Dilawar Husain, Yakub Ansari, and Imran Ahmad

Road Map of the Chapter

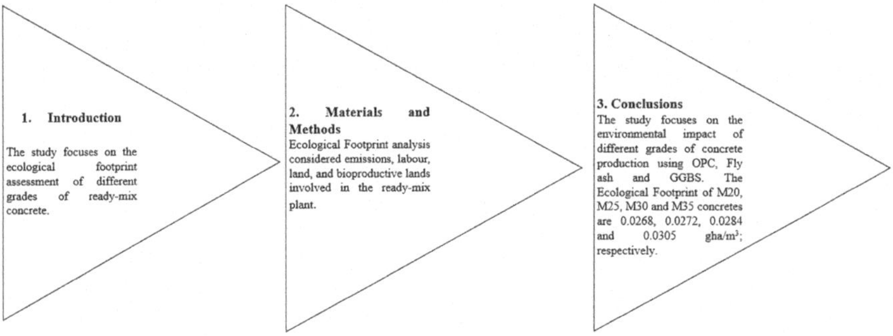

S. A. Mahevi · A. K. Kaliluthin
Department of Civil Engineering, School of Infrastructure, B.S. Abdur Rahman Cresent Institute of Science and Technology, Peerakankaranai, Tamil Nadu, India

D. Husain (✉)
Department of Mechanical Engineering, Maulana Mukhtar Ahmad Nadvi Technical Campus, Malegaon, Maharashtra, India

Y. Ansari · I. Ahmad
Department of Civil Engineering, Maulana Mukhtar Ahmad Nadvi Technical Campus, Malegaon, Maharashtra, India

S. S. Muthu (ed.), *Ecological Footprint of Industrial Spaces and Processes*, Environmental Footprints and Eco-design of Products and Processes,
https://doi.org/10.1007/978-3-031-69047-1_4

1 Introduction

Concrete is well-known and widely utilized among the most frequently used materials in modern constructions due to its imperviousness, tensile strength and affordability [1]. Primarily, it can be used in any climatic condition to build any type of structure [2]. About 32 million tons of concrete and 10% of cement dry composition amounting to a total of 4 billion tons are thought to be used annually [3]. Over the last four decades, the world's concrete consumption has more than doubled; meanwhile, per-capita consumption has reached around four tons [3]. It indicates that concrete production has a huge negative impact because it is accompanied by large emissions, waste generation and materials consumption [4, 5].

Ordinary Portland Cement (OPC) which is the main source of pollution as a raw material used in making concrete. Carbon dioxide emission of 657 kg per ton of cement production [1, 3]. The share of CO_2 emissions during cement production is shown in Fig. 1. It indicates that approximately 50% of the total CO_2 emissions related to cement production are generated during calcination preparation. Consequently, concrete production (consisting of cement) is highly concerned with mitigating negative environmental issues, mainly CO_2 emissions [6]. The construction industry has started taking measures to increase energy efficiency and use low-impact cementitious materials as a partial replacement for cement [7]. It may help to ensure sustainable development through using waste utilization, reuse and recycling. This would also make the environmental impacts of construction [7, 8]. The most feasible and sustainable way to reduce carbon emissions is by the usage of alternative cementitious materials (up to 50% replacement) in place of cement [8]. The details of embodied emission of different types of concrete raw materials are mentioned in Table 1.

The concrete is produced at the central batching plant and transported to the building site via heavy-duty truck-mounted Transit Mixers. In India, the ready-mix concrete industry is booming. In wealthy nations, ready-mix concrete makes up

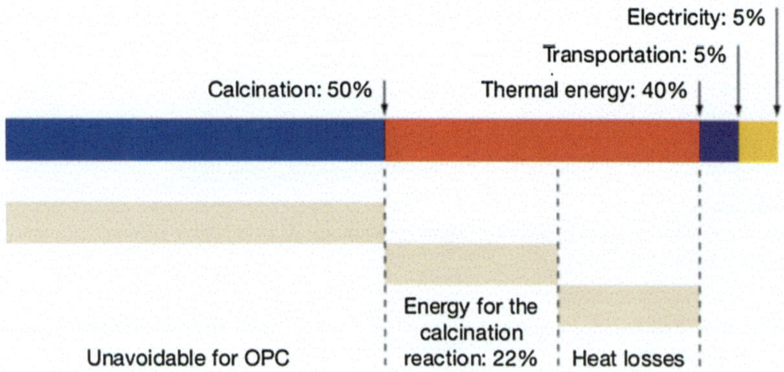

Fig. 1 Share of emissions (CO_2) during the cement production [6]

Table 1 The details of embodied emissions of different materials

S.No.	Materials	Embodied carbon ($kgCO_2/kg$) [9, 10]
1	Portland Cement	0.82–0.83
2	River Sand	0.0025–0.005
3	Crushed aggregate	0.0062
4	Fly ash	0.009–0.01
5	Slag	0.019–0.07
6	GGBS	0.067
7	Metakaolin	0.33
8	Silica fume	0.014
9	Water	0.0003
10	Superplasticizer	0.01–0.72

approximately 70% of cement use, whereas recast concrete makes up 25%. However, ready-mixed concrete makes up less than 5% of cement consumption in India, and site-mixed concrete makes up as much as 82%. In industrialized nations such as Japan, the ready-mix concrete industry consumes 70% of the cement produced, whereas in India, the industry uses only 2% of the total cement production.

Hussein et al. examined the factors contributing to the poor performance of ready-mixed concrete production plants, the study identified 23 significant reasons out of 35, with nine root causes responsible for 80% of the performance issues [11]. Broyles et al. experimentally examined the impact of concrete mixture proportions on the carbon footprint of concrete structures, utilizing data from Environmental Product Declarations (EPDs) of over 39,000 U.S.A. ready-mix concrete samples. Analysed using a custom Python script, the EPDs offer valuable insights into various environmental parameters such as Global Warming Potential, Ozone Depletion Potential, and Acidification Potential, among others [12]. Krisnawati et al. explore research delves into productivity enhancement strategies at XYZ Company, a construction materials producer in Banyumas, Indonesia, with a focus on production scheduling. By integrating product delivery systems and employing Genetic Algorithm techniques, the study develops efficient scheduling models. Results indicate a notable 24% reduction in job sequence completion times, demonstrating the potential for heightened productivity in the competitive construction industry [13]. Ghazi et al. These studies highlight plant classification, quality control systems, and statistical evaluations using SPSS software. Findings reveal that 75% of plants fail to meet the required criteria, prompting recommendations for improvement through the integration of influential variables and modern evaluation models [14].

Weerapura et al. proposed a novel digital twin (DT) system tailored for the ready-mix concrete (RMC) industry, aiming to mitigate production risks and enhance operational efficiency. The study assesses the technical feasibility, business value, and implementation challenges of DTs in the RMC industry [15]. Cordoba et al. investigated the environmental impact and material consumption of ready-mix concrete production in the Metropolitan Region of Buenos Aires, focusing on strategies to enhance sustainability. Material flow analysis and life cycle assessment

(LCA) were conducted, revealing that a significant portion of materials used consisted of aggregates, with Portland cement contributing the most to CO_2 emissions [16]. Delnavaz et al. explore sustainable methods to mitigate environmental impact in concrete production, focusing on the use of reclaimed water and supplementary cementitious materials like silica fume. Through a combination of mechanical testing, durability assessments, and life cycle analysis (LCA), the study evaluates the environmental and performance benefits of incorporating silica fume and reclaimed water into concrete mixtures. Results indicate that adding silica fume enhances environmental sustainability by reducing global warming potential (GWP) emissions, with potential reductions of up to 11.2% observed [17]. Senthamizh Sankar et al. investigated on internal and external factors influencing ready-mix concrete (RMC) plants, proposing risk control measures to enhance profitability. Through a structured approach involving risk identification, categorization, prioritization, and quantification, 12 high-consequence, high-probability risks are identified from 120 potential risks [18]. Fitriani et al. examined the implementation of a just-in-time (JIT) production system in the ready-mix concrete (RMC) industry to address issues like delays and quality concerns in raw material delivery. Findings indicate that while the concept of JIT hasn't been fully realized in RMC production due to challenges inherent in semi-manufacturing processes, there's potential for adaptation to enhance production productivity [19]. Lee et al. examined challenges in ready-mix concrete (RMC) management stemming from reliance on on-site experience rather than optimal planning. By synthesizing research on construction productivity simulations, the study develops a general expression to predict RMC placement time, enabling more efficient management. The simulation, applied to a Korean construction site, validates its effectiveness in optimizing RMC planning [20].

An extremely promising solution is based on the use of industrial waste, the inevitable byproduct of various industries, as a partial substitution for sand and cement, which, in turn, leads to the creation of materials with a smaller resource base and a smaller carbon footprint [1–3]. The recent findings show that the use of fly ash and GGBS can replace up to 30% of cement in concrete production while at the same time offering environmentally friendly and waste management solutions [21–23]. In the same way, water treatment sludge (WTS) has also been used as cement substitution in ready-mix concrete production [24]. The source of WTS is mostly from flocculation tanks, sedimentation tanks, and filter backwash wastewater containing minerals, sand particles, little organic matter, and coagulants [25]. WTS disposal is a challenge and could consequently lead to land requirements and secondary environmental issues [26]. WTS acts as a binder with which it is possible to partially replace (10–15% by weight of cement) cement without having much influence on strength and workability. [24].

Concrete with substitute cementitious materials needs to be assessed employing conventional environmental assessment tools to effectively understand the ecological footprint of such materials. Among numerous ones, the ecological footprint indicator is the environmental assessment tool to assess the impacts of products, processes, human activities etc.

The objectives of this study are to evaluate the ecological footprint assessment of different grades of ready-mix concrete (M20, M25, M30, M35 and M40 grades). The study also considers the possibility and feasibility of using fly ash, GGBS as a partial replacement of cement in concrete. The analysis considers the cumulative ecological footprint of energy consumption, raw material consumption, labour impact, and transportation impact of ready-mix concrete plants.

2 Materials and Methods

The study examines the environmental impact of the ready-mix plants using the Ecological Footprint tool. It commutatively incorporated raw materials consumption impact, energy used in the plant, transportation of materials and manpower. The study is more comprehensive than an assessment of the plant based on carbon footprint because carbon footprint analysis only considered emissions during the plant operation. However, Ecological Footprint analysis considered emissions, manpower, and all other bioproductive lands involved with a ready-mix plant.

2.1 Case Study

The case study ready-mix plant is situated in Nashik district of Maharashtra, India. The case study RMC plant consists of silos and bins for the storage of cement and aggregates respectively. For better mixing and high efficiency of material compositions (such as cement, sand aggregate, water, fly ash, GGBS etc.) of concrete, a computerized control system is used to control concrete production. The concrete production process and image of the case study ready-mix plant are shown in Fig. 2.

It has a capacity of the plant is 60 m^3 of concrete production per hour. The details of the ready-mix plant are mentioned in Table 2. The plant produced 30735 m^3 (M20: 10500 m^3; M25: 9405 m^3; M30: 7980 m^3 & M35: 2850 m^3) of all types of concrete in the year 2023. The annual raw materials consumption of the plant is mentioned in Table 3.

2.2 Ecological Footprint of Ready-Mix Concrete Plant

The annual Ecological Footprint of the case study pant is about 919.05 gha. In the year 2023, the annual M20, M25, M30 and M35 grade concrete production by the plant are 10500 m^3, 9900 m^3, 8400 m^3 and 3000 m^3; respectively. The material's environmental impact is the highest (719.27 gha) followed by transportation impact (170.66 gha). The details of all the parameters are as follows:

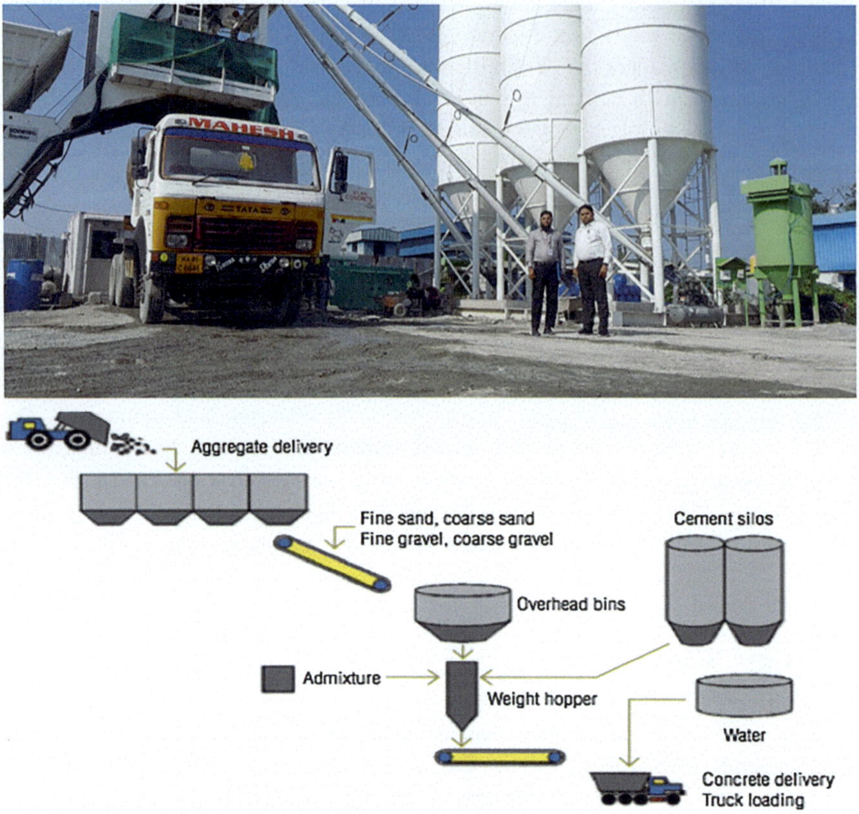

Fig. 2 The concrete production process and image of the case study ready-mix plant

(i) Ecological Footprint of Material's Impact

The environmental impact of raw materials consumption for concrete is significant. The impact depends on the types of materials, quantity, and transportation distance of raw materials. The section considers only materials types and quantities used for the ready-mix plant. The annual Ecological Footprint of materials consumption in the plant is 719.27 gha/year. The cement environmental contribution is highest (92.4%) among the rest of the materials used in the plant. The study suggested that cement replacement may reduce the total environmental impact of the plant. The details of annual materials consumption and its Ecological Footprint are mentioned in Table 4.

(ii) Ecological Footprint of Ready-Mix Machine Operation

The ready-mix concrete machine consumes 90 kWh of electricity for 60 m^3 of concrete production (i.e., 1.5 kWh/m^3 concrete production). The total annual electricity consumption is about 93015 kWh. The details of monthly electricity

Table 2 The details of the case study ready-mix plant

Sr. No	Description	Details
1	Manufacturer	RK Infra.
2	Capacity	60 m^3 concrete production per hour
3	Cement storage capacity	3 silos, 100-ton capacity/silo
4	Discharge height	4.1 m
5	Energy Consumption	90 kWh for 60 m^3 of concrete (Grid electricity) or 12 L diesel for 60 m^3 of concrete (Diesel generator)
6	Manpower requirement/year	9 FTE
7	Physical land	6 hectares
8	Transportation (from source to RMC plant)	
	Cement	80–120 km
	Sand	20–30 km
	Aggregate	20–30 km
	Fly Ash	80–100 km
	GGBS	80–100 km
	Admixture	05–10 km

Table 3 The details of raw materials consumption in the plant (in the year 2023)

Materials	Unit	Concrete grade				Total
		M20	M25	M30	M35	
Cement	ton	2856.0	2895.8	2664.9	1055.3	9471.9
Sand	ton	7207.2	5777.6	2.2	0.6	12987.6
Aggregate (10 mm)	ton	4764.4	4003.6	3284.0	1113.7	13165.6
Aggregate (20 mm)	ton	6898.5	5743.0	4665.8	1592.5	18899.7
Water	ton	21157.5	2128.5	1806.0	645.0	25737.0
Fly ash	ton	630.0	591.5	525.0	192.0	1938.5
GGBS	ton	157.5	74.3	144.9	54.0	430.7
Admixture	ton	24.5	25.1	23.3	9.2	82.1

consumption in shown in Fig. 3. The Ecological Footprint of Indian Grid electricity is reported as generation is 2.72×10^{-4} gha/kWh (ignoring factors other than emissions) [27]. The annual Ecological Footprint of ready-mix machine operation is estimated as 25.3 gha.

(iii) Ecological Footprint of Labour

The case study ready mix plant required 9 Full-time equivalent manpower for smooth operation. The plant operates 8 h/day for concrete production. The Ecological Footprint of Indian manpower is a range of $1.5–2.04 \times 10^{-4}$ gha/h [24]. The plant was operated 8 h/day and was opened for 300 days in the year 2023. The Ecological Footprint of labour impact is about 3.82 gha for the year 2023.

Table 4 The Ecological Footprint of different material consumption in the plant (in the year 2023)

Description	Unit	Quantity	Ecological Footprint/ unit	Ecological Footprint (gha)
Cement	kg	9471900	0.0000702	664.93
Sand	kg	12987581	0.000000157	2.04
Aggregate (size: 10 mm)	kg	13165590	0.000000776	10.22
Aggregate (size: 20 mm)	kg	18899745	0.000000776	14.67
Water	kg	25737000	7.28116E-08	1.87
Fly Ash	kg	1938525	3.31852E-06	6.43
GGBS	kg	430650	2.22341E-05	9.58
Admixture	kg	82136.25	0.000116148	9.54
Total				**719.27**

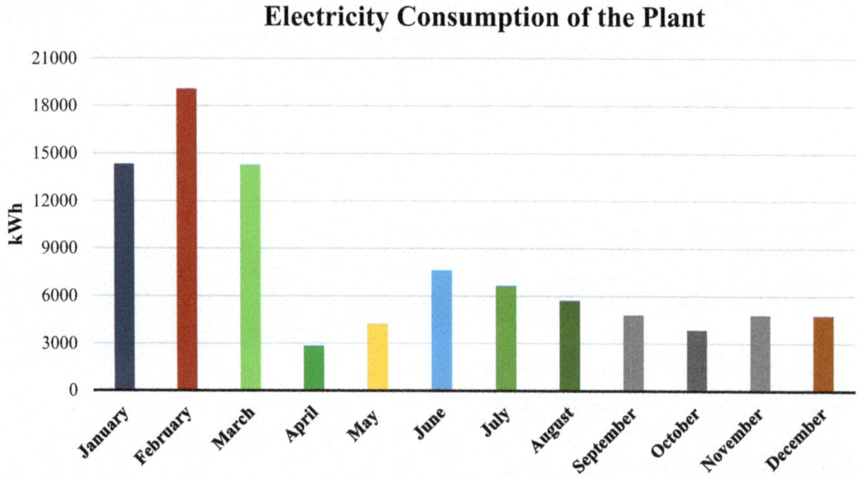

Fig. 3 Electricity consumption in the ready-mix plant

(iv) Ecological Footprint of Material Transportation

The Ecological Footprint of transportation is reported as 2.12–7.21 × 10^{-5} gha/ton-km [23, 28]. The transportation details of raw materials from its source to the case study plant are mentioned in Table 5. The Ecological Footprint of raw material's transportation impact of the case study plant is estimated as 170.66 gha for the year 2023.

Table 5 The transportation details of raw materials from its source to the case study plant

Description	Unit	Quantity	Distance (km)	Transportation ecological footprint/ unit	Average ecological footprint (gha)
Cement	kg	9471900	100–120	0.0000721 gha/ton-km	75.02
Sand	kg	12987581	20–30	0.0000721 gha/ton-km	23.375
Aggregate (size: 10 mm)	kg	13165590	20–30	0.0000721 gha/ton-km	23.7
Aggregate (size: 20 mm)	kg	18899745	20–30	0.0000721 gha/ton-km	34.02
Fly Ash	kg	1938525	80–90	0.0000721 gha/ton-km	11.865
GGBS	kg	430650	80–90	0.0000721 gha/ton-km	2.635
Admixture	kg	82136.25	5–10	0.0000721 gha/ton-km	0.045
Total					**170.66**

Table 6 The details of materials consumption in one m^3 of conventional concrete

Conventional Concrete per m^3					
		Concrete grade			
Materials	Unit	M20	M25	M30	M35
Cement	kg	320	360	380	400
Sand	kg	884	870	810	780
Aggregate (20 mm)	kg	685	730	695	700
Aggregate (10 mm)	kg	480	435	452	450
Water	kg	207	162	162	170.5

Table 7 The details of materials consumption in one m^3 of concrete using fly ash

Concrete (using fly ash) per m^3					
		Concrete grade			
Materials	Unit	M20	M25	M30	M35
Cement	kg	270	325	350	375
Fly ash	kg	60	75	85	70
Aggregate (20 mm)	kg	685	730	695	700
Aggregate (10 mm)	kg	480	435	452	450
Sand	kg	884	870	810	780
Water	kg	160	165	160	165
Admixture	kg	3.3	4.4	4.8	4.9

2.3 *Ecological Footprint of Concrete Production*

The study has estimated the Ecological Footprint of different grades of concrete. The Ecological Footprint of M20, M25, M30 and M35 concretes (consisting of OPC) are 0.029, 0.032, 0.033 and 0.035 gha/m^3; respectively. The details of materials consumption in one m^3 of different grades of concrete are mentioned in Tables 6, 7, and 8. The details of the Ecological Footprint of different grades (with or without partial replacement of cement) of concrete are shown in Fig. 4.

Table 8 The details of materials consumption in one m^3 of concrete using fly ash and GGBS

Concrete (with Fly ash + GGBS) per m^3					
		Concrete grade			
Materials	Unit	M20	M25	M30	M35
Cement	kg	180	200	220	250
Fly ash	kg	40	50	50	60
GGBS	kg	100	100	115	120
Aggregate (20 mm)	kg	722	713	701	691
Aggregate (10 mm)	kg	451	462	466	463
Sand	kg	730	718	703	684
Water	kg	215	214	216	210
Admixture	kg	3.04	3.5	3.85	4.3

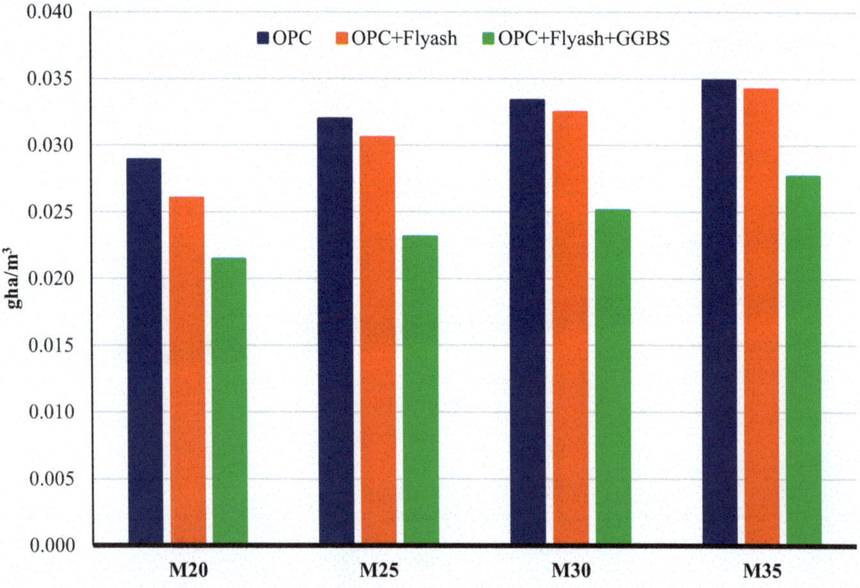

Fig. 4 Ecological Footprint of different grade of concretes

The study also examines the potential reduction of the Ecological Footprint of concretes using fly ash and GGBS as a partial replacement of cement in different grades of concrete. The study indicates that up to 25.7% of bioproductive land demand may be reduced by using fly ash (12.5% by weight) and GGBS (31.25% by weight) as a partial replacement of cement in M20 concrete without compromising the compressive strength of the concrete.

3 Conclusions

The study focuses on the environmental impact of different grades of concrete production using OPC, Fly ash and GGBS. The Ecological Footprint of M20, M25, M30 and M35 concretes are 0.029, 0.032, 0.033 and 0.035 gha/m^3; respectively. The study suggested that the Ecological Footprint of materials used for concrete production has the highest among all other parameters like materials transportation, energy consumption, labour etc. Concrete production using the combination of fly ash and GGBS (as a partial replacement of OPC) has a lower environmental impact than conventional concrete (consisting of OPC) production.

Ready-mix concrete production involves significant amounts of raw materials and bioproductive lands that indirectly harm the planet. The energy consumption associated with the transportation of ready-mix concrete production is the second most environmental concern after raw materials consumption. The operation of the plants, as well as the transportation of concrete mixers to construction sites, depends heavily on fossil fuels and is also responsible for greenhouse gas emissions.

References

1. York, I. N., & Europe, I. (2021). Concrete needs to lose its colossal carbon footprint. *Nature, 597*(7878), 593–594.
2. Miller, S. A., Horvath, A., & Monteiro, P. J. (2016). Readily implementable techniques can cut annual CO_2 emissions from the production of concrete by over 20%. *Environmental Research Letters, 11*(7), 074029.
3. Habert, G., Miller, S. A., John, V. M., Provis, J. L., Favier, A., Horvath, A., & Scrivener, K. L. (2020). Environmental impacts and decarbonization strategies in the cement and concrete industries. *Nature Reviews Earth & Environment, 1*(11), 559–573.
4. Monteiro, P. J., Miller, S. A., & Horvath, A. (2017). Towards sustainable concrete. *Nature Materials, 16*(7), 698–699.
5. Ansari, Y., Husain, D., Haadi, S. M., Haloi, J., & Prakash, R. (2023). Life cycle ecological footprint of building: A case study of low-rise tropical residential building. *International Journal of Environmental Science and Technology, 20*(9), 9779–9794.
6. Lowitt, S. (2020). *Towards the decarbonisation of the south African cement industry: Opportunities and challenges*. Trade and Industrial Policy Strategies (TIPS).
7. Schiller, G., Gruhler, K., & Ortlepp, R. (2017). Continuous material flow analysis approach for bulk nonmetallic mineral building materials applied to the German building sector. *Journal of Industrial Ecology, 21*(3), 673–688.
8. Cui, K., Liang, K., Chang, J., & Lau, D. (2022). Investigation of the macro performance, mechanism, and durability of multiscale steel fiber reinforced low-carbon ecological UHPC. *Construction and Building Materials, 327*, 126921.
9. Dobiszewska, M., Bagcal, O., Beycioğlu, A., Goulias, D., Köksal, F., Płomiński, B., & Ürünveren, H. (2023). Utilization of rock dust as cement replacement in cement composites: An alternative approach to sustainable mortar and concrete productions. *Journal of Building Engineering, 69*, 106180.
10. Zeybek, Ö., Özkılıç, Y. O., Karalar, M., Çelik, A. İ., Qaidi, S., Ahmad, J., et al. (2022). Influence of replacing cement with waste glass on mechanical properties of concrete. *Materials, 15*(21), 7513.

11. Hussein, S. G., & Al-Zwainy, F. M. (2024). Diagnosing and identifying standards affecting on the ready-mix concrete production plants performance: An analytical study. *Tikrit Journal of Engineering Sciences, 31*(1), 211–222.
12. Broyles, J. M., Gevaudan, J. P., & Brown, N. C. (2024). A compiled dataset of ready-mix concrete environmental product declarations for life cycle assessment. *Data in Brief, 52*, 109852.
13. Krisnawati, M., Syailendri, F. R. A., & Sibarani, A. A. (2024). Ready mix concrete production scheduling and truck mixer allocation using genetic algorithm: A case study. *Engineering Headway, 6*, 303–311.
14. Ghazi, S., Al-Zwainy, F. M. S., & Manogaran, G. M. (2023). The critical review to evaluate performance of ready-mix concrete production plant: Critical review to evaluate performance of concrete plant. *Al-Nahrain Journal for Engineering Sciences, 26*(3), 205–215.
15. Weerapura, V., Sugathadasa, R., De Silva, M. M., Nielsen, I., & Thibbotuwawa, A. (2023). Feasibility of digital twins to manage the operational risks in the production of a ready-mix concrete plant. *Buildings, 13*(2), 447.
16. Cordoba, G., Paulo, C. I., & Irassar, E. F. (2023). Towards an eco-efficient ready mix-concrete industry: Advances and opportunities. A study of the Metropolitan Region of Buenos Aires. *Journal of Building Engineering, 63*, 105449.
17. Delnavaz, M., Sahraei, A., Delnavaz, A., Farokhzad, R., Amiri, S., & Bozorgmehrnia, S. (2022). Production of concrete using reclaimed water from a ready-mix concrete batching plant: Life cycle assessment (LCA), mechanical and durability properties. *Journal of Building Engineering, 45*, 103560.
18. Senthamizh Sankar, S., Anandh, K. S., Raja Pandian, S., & Aravinth, K. S. (2022). Investigating the ways to optimize the production of ready mix concrete plants. In *Advances in Construction Management: Select Proceedings of ACMM 2021* (pp. 369–380). Springer Nature Singapore.
19. Fitriani, H., & Rizki, L. D. (2022). Just-in-time application on readymix concrete production. *International Journal of Advanced Technology and Engineering Exploration, 9*(93), 1183.
20. Lee, S. J., Jung, K. T., Youn, J. Y., & Lee, D. (2022). Optimal management plan through ready-mix concrete placement and transportation time. *International Journal of Mechanical Engineering, 6*(3), 866–872.
21. Adesina, A. (2020). Recent advances in the concrete industry to reduce its carbon dioxide emissions. *Environmental Challenges, 1*, 100004.
22. El-Hassan, H., & Kianmehr, P. (2017). Sustainability assessment and physical characterization of pervious concrete pavement made with GGBS. In *MATEC web of conferences* (Vol. 120, p. 07001). EDP Sciences.
23. Husain, D., & Prakash, R. (2019). Ecological footprint reduction of built envelope in India. *Journal of Building Engineering, 21*, 278–286.
24. Ansari, Y., Husain, D., Das, U. K., Haloi, J., Khan, N. A., Prakash, R., & Husain, M. (2023). Ecological footprint assessment of concrete: Partial replacement of cement by water treatment sludge and stone dust. *Sustainability, 15*(9), 7512.
25. Ahmad, T., Ahmad, K., & Alam, M. (2016). Sustainable management of water treatment sludge through 3 'R'concept. *Journal of Cleaner Production, 124*, 1–13.
26. Zhao, Y., Nzihou, A., Ren, B., Lyczko, N., Shen, C., Kang, C., & Ji, B. (2021). Waterworks sludge: an underrated material for beneficial reuse in water and environmental engineering. *Waste and Biomass Valorization, 12*, 4239–4251.
27. Biswas, A., Husain, D., & Prakash, R. (2021). Life-cycle ecological footprint assessment of grid-connected rooftop solar PV system. *International Journal of Sustainable Engineering, 14*(3), 529–538. https://doi.org/10.1080/19397038.2020.1783719
28. Nawandar, V., Husain, D., & Prakash, R. (2021). Ecological footprint assessment and its reduction for packaging industry. In S. S. Muthu (Ed.), *Assessment of ecological footprints. Environmental footprints and eco-design of products and processes*. Springer. https://doi.org/10.1007/978-981-16-0096-8_2

Quantifying the Environmental Footprint: An In-Depth Assessment of the Environmental Impact Associated with the Production of California Red Earthworm Meal (CRWM)

Yhoan Sebastian Gaviria Gaviria and José Edgar Zapata Montoya

1 Roadmap of the Chapter

In Fig. 1, the roadmap of the chapter is shown, identifying each of the stages and the main parts of these stages. The chapter begins with an introduction about the generation of organic waste and the problems it causes. Subsequently, it discusses the methodology for quantifying environmental impact through ecological footprint analysis. Finally, it explores an alternative for the use of organic waste through a biotechnology such as vermiculture. The next section is the methodology of the study, starting with the system delineation, the mathematical models for environmental quantification, and the characterization of the paste obtained at the end of the process. Next, the results of the study are presented, comparing and discussing them with other similar studies, to finally conclude.

2 Introduction

The economic growth of many nations worldwide is connected to the depletion of natural resources. The current human demand for water, food, infrastructure, and energy puts ecological pressure on the planet, leading to various environmental deteriorations such as the reduction of natural resources, waste generation, and greenhouse gas emissions [1]. The utilization of these resources gradually increases

Y. S. G. Gaviria (✉) · J. E. Zapata Montoya
Nutrition and Food Technology Research Group, NUTEC, Universidad de Antioquia, Medellín, Colombia
e-mail: Yhoan.gaviria@udea.edu.co; edgar.zapata@udea.edu.co

S. S. Muthu (ed.), *Ecological Footprint of Industrial Spaces and Processes*, Environmental Footprints and Eco-design of Products and Processes,
https://doi.org/10.1007/978-3-031-69047-1_5

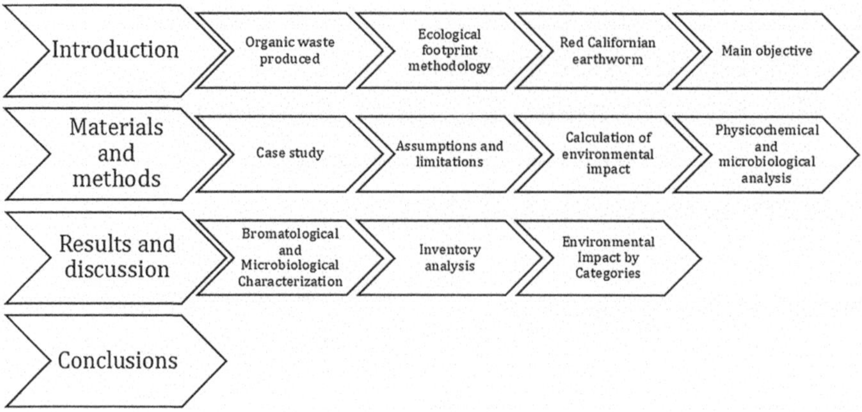

Fig. 1 Roadmap of the chapter

to a level that causes environmental degradation. Consequently, the availability of natural resources (biocapacity) becomes limited due to the insufficient capacity to mitigate the impacts of waste generation [2]. In this context, there has been a growing necessity for the implementation of energy and environmental policies aimed at preserving natural resources [2, 3], which has led to the development of various methodologies to control environmental degradation by assessing the impact of different processes. Among these, the ecological footprint stands out as a methodology capable of providing a comprehensive perspective on the sustainability of a process or even a country [4].

The ecological footprint is a methodology that evaluates the impact of human activities based on land use, water consumption, forest products, infrastructure, and carbon emissions [5]. Currently, this methodology is regarded as one of the most extensively used sustainability indicators because it provides comparable, reliable, and comprehensive results [1, 6]. It assesses the effects of human activities on the environment by calculating the amount of global hectares of land and water required to produce the resources and absorb the waste generated [1]. Consequently, the ecological footprint is employed in environmental impact assessments within optimization processes, enabling the analysis of each stage, identifying the activities that generate the most significant impact, and implementing strategies to reduce and enhance the ecological performance of the process [7].

Different species of earthworms exhibit detritivorous capabilities, facilitated by their digestive enzymes and the aerobic and anaerobic microflora in their gut, which enable the biodegradation of organic waste from agricultural and agro-industrial activities [8]. This process, known as vermicomposting, produces two products of high biological and commercial value: compost, derived from the decomposition of waste from various industries, and worm protein, obtained from the prolific reproduction of the red Californian earthworm [9, 10]. Earthworms are classified in the animal kingdom as terrestrial annelids within the class Oligochaeta. They thrive in moist environments and feed on decomposing plant and animal organic matter,

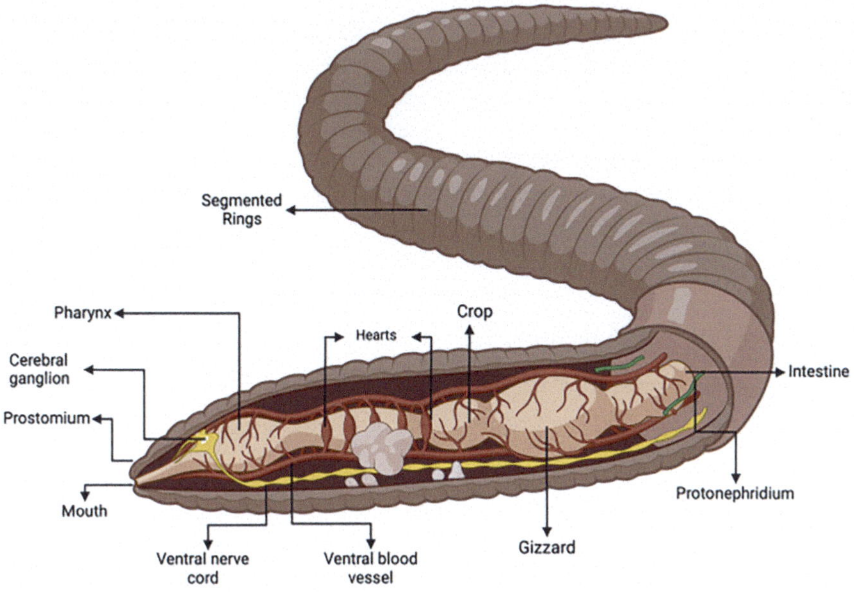

Fig. 2 Morphology of the Red Californian Earthworm (*Eisenia fetida*)

making them excellent recyclers [11]. There are approximately 8000 known species of earthworms, but only 3500 have been studied and classified, indicating that not all species are suitable for breeding due to their specific and challenging habitat requirements. However, the species *Eisenia fetida*, commonly known as the red Californian worm, is exceptionally well-adapted for captivity [12]. The body of annelids is characterized by both external and internal segmentation. The body wall consists of several layers (Fig. 2): an outer cuticle (a thin, generally colorless layer), an epidermis (a simple epithelium with glandular cells that secrete mucus or serum), muscle layers (an outer circular layer and an inner longitudinal layer), the peritoneum (an innermost layer that borders the worm's coelom externally), and the coelom (a cavity containing coelomic fluid that extends throughout the animal and surrounds the alimentary canal) [13].

These animals are photophobic, meaning they cannot tolerate ultraviolet sunlight. Exposure to sunlight can result in dehydration and death. A moist environment is crucial for their survival, but they cannot endure overly watery conditions since they breathe through their skin [14]. Earthworms are hermaphroditic and among the few species capable of asexual reproduction, meaning they do not require two individuals to reproduce [15]. Their sense of direction is limited, as they lack eyes and ears; their most developed senses are touch and taste, which help them locate food and evade predators. Additionally, their movement is regulated by light-sensitive cells in their skin [11, 15, 16].

The Californian Red Worm (Eisenia fetida) is the most widely used species globally for producing humus (organic fertilizer) [17]. Its importance lies not only in its ability to efficiently convert organic waste into nutrient-rich material but also in its high protein content [18]. Due to this, the Californian Red Worm has been utilized as a substitute for protein-rich raw materials in animal feed, as it is nutritionally suitable for various species [19], such as cattle [20], poultry [21, 22], and seafood [23, 24]. Thus, the objective of this chapter was to determine the environmental impact generated in the process of obtaining Californian Red Worm Paste (*Eisenia fetida*) (CRWP), using the ecological footprint methodology as a sustainability indicator.

3 Materials and Methods

3.1 Case Study

The production system for obtaining Californian Red Worm Paste (CRWP) is outlined in Fig. 3, detailing the five stages and the inputs and outputs at each stage. The first stage involves the cultivation and breeding of the Californian Red Worm, with an annual production base of 46.56 kg per square meter of worms. The feeding substrate comprised a mixture of organic waste and cattle manure in a 1:3 ratio, respectively. The worms were manually separated from the substrate and washed with drinking water to remove any remaining substrate residues. They were then purged for 4 h in a 4% sodium bicarbonate ($NaHCO_3$) solution by immersion in a plastic container to eliminate any internal substrate residues. Subsequently, the worms were sacrificed by immersion in a 7% saline solution (NaCl) for 30 min in a plastic container. Finally, the worms were rinsed with drinking water and stored at −20 °C for future use. Measurements of the quantities of each required or generated input and output were conducted at each stage of the process.

3.2 Assumptions and Limitations

For the present study, certain assumptions were made, including:

- The impact of buildings was not considered in the calculations due to their long lifespan.
- The environmental impact of industrial equipment was not included in the environmental analysis due to limitations in obtaining data on manufacturing

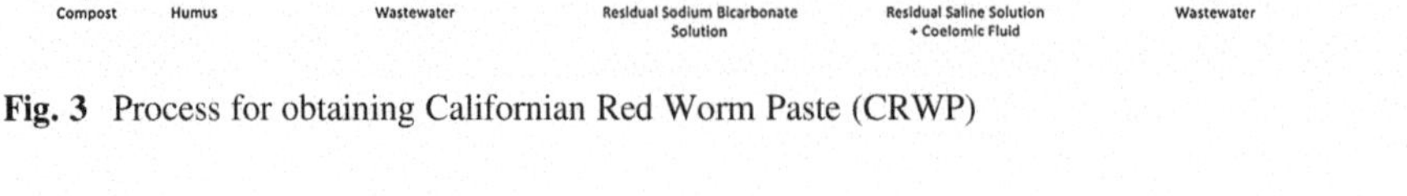

Fig. 3 Process for obtaining Californian Red Worm Paste (CRWP)

processes. However, it has been demonstrated that these do not significantly affect the results of the environmental impact assessment because they have long lifespans [25, 26].

3.3 Calculation of Environmental Impact

To analyze the environmental impact of the process of obtaining Californian Red Worm Paste, the ecological footprint study was utilized. This methodology followed the standards set by Gwehenberger and Narodoslawsky [7] and Krotscheck and Narodoslawsky [27], which employ calculation models to estimate the ecological impact of each activity depicted in Fig. 2. This approach enhances the accuracy of the calculations as it requires extensive data collection and simultaneously reduces the limitations associated with using international databases [28]. The environmental impact values of the inputs and outputs at each stage of the process were normalized to a single unit (Ha/ton), representing the amount of forest hectares needed to sequester the CO_2 produced during the process.

Equation (1) was utilized to quantify the environmental impact (EI) resulting from the organic compounds (co) utilized or generated within the process. This calculation adhered to the guidelines established by the Intergovernmental Panel on Climate Change (IPCC). Within the equation, CRO represents the amount of organic waste, while FCM denotes the correction factor for methane gas (CH4), dependent on the specific solid waste management practices within a given sector [29]. COD signifies the degradable organic carbon fraction of the waste that can undergo biochemical decomposition [30], whereas CODF refers to the fraction of organic carbon not assimilated or undergoing very slow degradation [29]. Additionally, F and R represent the fractions of CH4 within landfill gas and the CH4 recovered, respectively [31], The oxidation factor of CH4 is indicated as OX, and the global warming potential of methane (GWP) for a 100-year timeframe was incorporated. Furthermore, FF stands for the CO_2 fixation factor specific to the region defined within the environmental analysis [32].

$$EI = \sum_{OC=1}^{n} CRO\left(\frac{16 * (FCM * DCO * COD_F * (F - R) * (1 - OX) * GWP)}{12 * FF}\right) \quad (1)$$

The environmental impact stemming from the utilization of electrical energy across various stages was determined using Eq. (2). In this equation, EEG represents the consumption of electrical energy e within each stage of the process, while EEE signifies the CO_2 emissions factor attributed to one kWh of energy.

$$EI = \sum_{e=1}^{n} EE_G * \frac{EEE}{FF} \tag{2}$$

In the washing subsystems, it's essential to incorporate an input flow representing the cleaning water. To assess the environmental impact of water supply, Eq. (3) was utilized. Here, WL denotes the volume of water required in the process, while EBA represents the embedded energy for water supply.

However, the utilization of water entails an environmental impact upon its disposal. This impact was evaluated using Eq. (4), wherein RWL signifies the volume of residual water generated during the cleaning process, CODBO5 represents the biochemical oxygen demand of the degradable fraction within the residual water [29], and CFMCH4 denotes the maximum methane production capacity of said fraction.

$$EI = \sum_{a=1}^{n} W_L \left(\frac{EB_A * EEE}{FF} \right) \tag{3}$$

$$EI = \sum_{a=1}^{n} RW_L * \left(\frac{CO_{DBO5} * MPC_{CH4} * CFM_{CH4} * PCG}{FF} \right) \tag{4}$$

To assess the environmental impacts linked with production ingredients and other input flows across various subsystems, Eq. (5) is employed. Within this equation, RQI represents the necessary quantity of product i, EBPI corresponds to the embedded energy for its manufacturing, and CI denotes the maximum load in product transportation.

$$EI = RQ_I \left(\frac{EB_{PI} * EEE + \left(\frac{D_R}{R_C * C_I} \right) * EC}{FF} \right) \tag{5}$$

3.4 Physicochemical and Microbiological Analysis

The proximate characterization of Californian Red Worm Paste (CRWP) was conducted in compliance with the guidelines outlined by the Association of Official Analytical Chemists (AOAC) [33]. The moisture content was determined following standard 930.15, involving the drying of the sample at 105 °C for 12 h. The protein percentage was assessed using the Kjeldahl method (AOAC 954.010), while the total ash content was determined in accordance with AOAC 942.05. The ether extract content (total lipid content) was measured using the Soxhlet methodology as per AOAC 920.39 [34].

Furthermore, the quantification of protein types (albumins, globulins, glutelins, and prolamins) was carried out using the Osborne method. Microbiological characterization adhered to the parameters established by the Colombian Technical Standard NTC 4519. This standard delineates the analysis of aerobic mesophiles, total and fecal coliforms (E. coli), Salmonella spp., and molds and yeasts for this type of product [35].

4 Results and Discussion

4.1 Bromatological and Microbiological Characterization

Table 1 presents the proximate composition values of the paste derived from the Californian Red Worm (*Eisenia fetida*). It is notable that the paste exhibits a notably high protein content compared to other macro compounds, primarily composed of albumin, globulin, and glutelin. This protein concentration falls within the typical range reported in the literature, indicating that earthworms typically possess protein values exceeding 60% [36, 37]. In terms of ash content, the observed values were slightly higher than those reported by Garczyńska for the earthworm (*Dendrobaena veneta Rosa*); however, the fat content is markedly lower in comparison (12.21%) [36]. The substantial proportion of proteins with high biological value renders this substrate a promising alternative for various biotechnological applications such as enzymatic hydrolysis [17, 37].

On the other hand, Table 2 illustrates the microbiological characterization of Californian Red Worm Paste obtained for this study, emphasizing the absence of pathogenic microorganisms such as fecal coliforms, *Salmonella* spp., and *Clostridium* sulfite-reducing spores. Additionally, it showcases the low concentration of aerobic mesophilic bacteria, molds and yeasts, and total coliforms, all of which adhere to the parameters established by current regulations. These findings suggest that the purging and cleaning process of the worm paste is effective, yielding a product with suitable characteristics for implementation in processes intended for human consumption.

Table 1 Bromatological composition of Californian Red Worm Paste

Component	CRWM (%)
Moisture (%)	84.87 ± 0.06
Protein (%)[a]	67.06 ± 0.08
Fat (%)[b]	1.42 ± 0.38
Ash (%)[b]	5.82 ± 0.39
Albumins (%)	65.25 ± 0.28
Globulins (%)	20.12 ± 0.12
Prolamins (%)	2.85 ± 0.07
Glutelins (%)	11.87 ± 0.09

[a]Dry base
[b]Moisture base

Table 2 Microbiological composition of Californian Red Worm Paste

Analysis	Value	Requirements[a]
Aerobic mesophiles (ufc/ml)	5×10^2	10×10^5
Total coliforms (ufc/ml)	3.5×10^1	10×10^4
Fecal coliforms	Absence	Absence
Salmonella spp	Absence	Absence
Clostridium sulfite spore	Absence	Absence
Molds and yeasts (ufc/ml)	8.2×10^1	10×10^4

[a]Directiva técnica DIP 30-100-001; Instituto Colombiano de Agricultura (ICA)

4.2 Inventory Analysis

The material and energy balance in the production process of obtaining Californian Red Worm Paste was conducted through inventory analysis (refer to Table 3), which meticulously outlines the inputs and outputs at each stage of the process (as depicted in Fig. 3) along with their designated quantities based on a calculation basis of 46.56 kg per m^2. Furthermore, the environmental impact (EI) in $kgCO_2$ per year is observed.

Subsystem 1, corresponding to the breeding of Californian Red Worms, exhibits a reduction in environmental impact of approximately −3913 $kgCO_2$/year attributed to the utilization of organic waste and manure for worm feeding. It's noteworthy that this value carries a negative sign, signifying that it represents a byproduct of agroindustry that, if left unused, would generate an environmental impact of equivalent magnitude. Therefore, by employing it for worm breeding, this adverse impact on the environment is circumvented. Similar trends and values of environmental impact have been reported by Malakahmad et al. [38] across various processes of organic waste disposal, wherein the avoided CO_2 emissions from anaerobic digestion notably exceeded the total CO_2 emissions of the process, leading to negative values. Moreover, at this stage, two products with substantial economic value are generated: liquid humus and vermicompost.

From the breeding stage, a production of 46.56 kg of fresh worms is obtained, which then enter the process of obtaining paste or protein. In the two washing stages, drinking water is utilized, which subsequently exits the system with a heightened organic load, as evidenced by the rise in environmental impact due to the increase in BOD_5. The total yield of the Californian Red Worm Paste production process amounts to 94.5%. The primary reduction (1.4 kg) occurs during the stage of sacrifice with saline solution. This reduction is attributed to the excretion of coelomic fluid by the worms under low-stress conditions. This fluid is produced by specialized cells known as coelomocytes, which primarily contribute to the worm's immunity [39].

Table 3 Inventory analysis of Californian Red Worm Paste production

Cultivation worms		
Input	**Quantity**	**EI (kgCO_2/year)**
Breeding Stock (kg)	3	0
Organic waste (kg)	300	−3880.8
Water (m^3)	0.36	0.165
Cattle manure (kg)	900	−32.625
Output	**Quantity**	**EI (kgCO_2/year)**
Compost (kg)	1000	0
Humus (L)	10	0
Worm (kg)	46.56	−3913.259
Initial washing		
Input	**Quantity**	**EI (kgCO_2/year)**
Worm (kg)	46.56	−3913.259
Drinking water (m^3)	0.6	0.276
Output	**Quantity**	**EI (kgCO_2/year)**
Wastewater (m^3)	0.6	2.52
Washed worm (kg)	46.094	−3910.463
Purge		
Input	**Quantity**	**EI (kgCO_2/year)**
Washed worm (kg)	46.094	−3910.463
Sodium Bicarbonate Solution 4% (m^3)	57.618	5.146
Output	**Quantity**	**EI (kgCO_2/year)**
Residual Sodium Bicarbonate Solution (m^3)	58.078	87.118
Purged worm	45.633	−3818.198
Sacrifice		
Input	**Quantity**	**EI (kgCO_2/year)**
Purged worm (kg)	45.633	−3818.198
Saline solution 7% (m^3)	57.041	2.1059
Output	**Quantity**	**EI (kgCO_2/year)**
Residual saline solution (m^3)	58.410	87.616
Dead worm (kg)	44.264	−3728.476
Final washing		
Input	**Quantity**	**EI (kgCO_2/year)**
Dead worm (kg)	44.264	−3728.476
Water (m^3)	0.3	0.138
Output	**Quantity**	**EI (kgCO_2/year)**
Wastewater (m^3)	0.3	1.26
Meal worm (kg)	44	−3727.077

EI Environmental impact

4.3 Environmental Impact by Categories

The primary impact categories of the production system are depicted in Fig. 4, highlighting the key sources of greenhouse gas emissions. Within these impact categories, similar inputs and outputs were consolidated. The impact of water considers emissions from both the input water for cleaning and its final disposal (wastewater). Despite the fact that the process of obtaining compost and Californian Red Worm Paste results in a reduction in environmental impact by utilizing waste, it's notable that CO_2 emissions are also generated during the process. The disposal of wastewater from washing emerges as the primary cause of these emissions (4.1 $kgCO_2$/(kg year)), followed by chemical inputs ($NaHCO_3$, NaCl) (0.16 $kgCO_2$/(kg year)), and lastly, the production of drinking water (0.013 $kgCO_2$/(kg year)).

Regarding the wastewater disposal process, its effects are primarily attributed to the organic load present from the worm feed substrate residues. This organic load complicates the final disposal process as it necessitates more oxygen for degradation, thereby diminishing its capacity to assimilate pollutant load and naturally restore its quality [40]. Similar findings have been reported in assessments of manure utilization for vermicomposting, indicating that uncontrolled discharge of animal manure into water bodies directly contributes to their eutrophication [41]. The utilization of organic waste from both agro-food (organic material) and livestock (manure) industries not only facilitates a direct reduction in environmental impact but also generates added value to by-products, thereby impacting the process economy positively [42]. Vermicomposting emerges as a viable method of manure management,

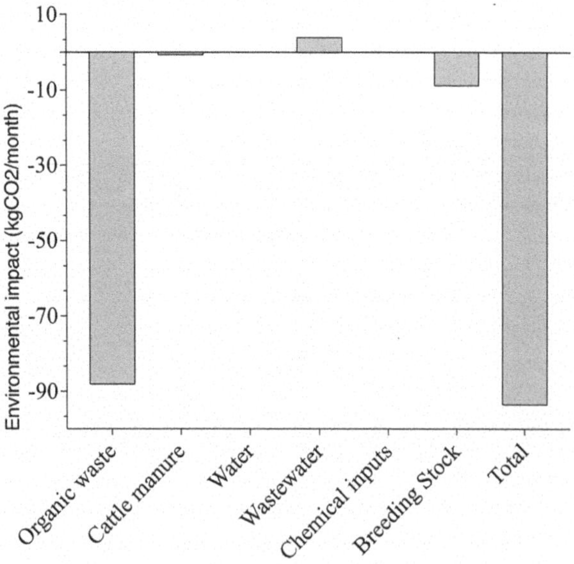

Fig. 4 Environmental impact by categories in the production of Californian Red Worm Paste (CRWP)

reducing waste by 45.9% with a waste-to-biomass conversion rate of 3.5%, as reported by Singh and Singh [43]. It has been observed that vermicomposting systems exhibit better environmental performance primarily due to low ammonia (NH_4) emissions both in their obtaining process and application [44].

5 Conclusions

The utilization of ecological footprint as a methodology for environmental assessment in the production process of Californian Red Worm Paste (*Eisenia fetida*) enabled the quantification of the genuine environmental impact associated with the utilization of waste from agro-food and livestock industries. This approach facilitated the identification of stages and input-output flows with the highest CO_2 emissions, offering valuable insights for the implementation of new alternative technologies aimed at achieving both technological and environmental sustainability.

Moreover, vermiculture demonstrates high feasibility as it yields two products of significant biological and economic value: vermicompost and worm paste. This not only reduces environmental impact through the utilization of agro-industrial waste but also underscores the potential for sustainable resource management and circular economy practices.

References

1. Ahmed, Z., & Wang, Z. (2019). Investigating the impact of human capital on the ecological footprint in India: An empirical analysis. *Environmental Science and Pollution Research, 26*, 26782–26796.
2. Sharif, A., Baris-Tuzemen, O., Uzuner, G., Ozturk, I., & Sinha, A. (2020). Revisiting the role of renewable and non-renewable energy consumption on Turkey's ecological footprint: Evidence from Quantile ARDL approach. *Sustainable Cities and Society, 57*, 102138.
3. Ibidhi, R., Hoekstra, A. Y., Gerbens-Leenes, P. W., & Chouchane, H. (2017). Water, land and carbon footprints of sheep and chicken meat produced in Tunisia under different farming systems. *Ecological Indicators, 77*, 304–313.
4. Syrovátka, M. (2020). On sustainability interpretations of the Ecological Footprint. *Ecological Economics, 169*, 106543.
5. Ahmed, Z., Asghar, M. M., Malik, M. N., & Nawaz, K. (2020). Moving towards a sustainable environment: The dynamic linkage between natural resources, human capital, urbanization, economic growth, and ecological footprint in China [Forthcoming]. *Resources Policy, 67*, 101677.
6. Jóhannesson, S. E., Davíðsdóttir, B., & Heinonen, J. T. (2018). Standard ecological footprint method for small, highly specialized economies. *Ecological Economics, 146*, 370–380.
7. Gwehenberger, G., & Narodoslawsky, M. (2007). The ecological impact of the sugar sector-Aspects of the change of a key industrial sector in Europe. *Computer Aided Chemical Engineering, 24*, 1029–1034.

8. Yuvaraj, A., et al. (2021). Activation of biochar through exoenzymes prompted by earthworms for vermibiochar production: A viable resource recovery option for heavy metal contaminated soils and water. *Chemosphere, 278.*
9. Muñoz-Rojas, M., Delgado-Baquerizo, M., & Lucas-Borja, M. E. (2021). Soil biodiversity and organic carbon are essential to reverse desertification. *Ecosistemas, 30* Preprint at https://doi.org/10.7818/ECOS.2238
10. Riascos-Vallejos, A. R., Crespo-López, G., & Medina-Mesa, Y. (2022). Effect of food source on the chemical composition of Californian Red Worm (Eisenia foetida) vermicompost. *Cuban Journal of Agricultural Science, 56*, 191–200.
11. Çinar, K., Öztop, M., & Demirbağ, E. (2014). The histochemical characterization of the glycoconjugates in the epidermal mucous cells of the Red Californian Earthworm, Eisenia foetida. *Journal of Histology, 2014*, 1–6.
12. Reines, M. (1998). *Lombrices de Tierra Con Valor Comercial: Biología y Técnicas de Cultivo.*
13. Berman, W. (1986). *Exploring with probe and scalped. How to dissect special projects for advanced study.* Hall Press.
14. Schiedeck, G., Gonçalves, M. M., & Schwengber, J. E. (2006). *Minhoculture and production of humus for family farming*. Circular Técnico.
15. Edwards, C. A., & Lofty, J. R. (2013). *Biology of earthworms.* (Originally published by Chapman and Hall Ltd). https://doi.org/10.1007/978-1-4899-6912-5
16. Edwards, C. A., Hendrix, P. F., Arancon, N. Q., & Dumanig, F. (1996). *Biology and ecology of earthworms*. Springer.
17. Muller, M., et al. (2017). Study of different conditions of enzymatic hydrolysis in earthworms. *International Journal of New Technology and Research, 3*, 263202.
18. Guerrero, R. D. (2009). Earthworm culture for vermicompost and vermimeal production and for vermiceutical application in the Philippines (1978–2008) – A review. *Global Science Books, 3*, 89–97.
19. Oanh, H. N. (2016). Survey process of autolysis from earthworm Perionyx Excavates obtain protein solution. *MOJ Proteomics & Bioinformatics, 3*, 48–50.
20. Moreki, J., & Tiroesele, B. (2012). Termites and earthworms as potential alternative sources of protein for poultry. *International Journal for Agro Veterinary and Medical Sciences, 6*, 368.
21. Sinha, R. K., Agarwal, S., Chauhan, K., Chandran, V., & Soni, B. K. (2010). Vermiculture technology: Reviving the dreams of Sir Charles Darwin for scientific use of earthworms in sustainable development programs. *Technology Investment, 01*, 155–172.
22. Sinha, R. K., Valani, D., Chauhan, K., & Agarwal, S. (2010). Embarking on a second green revolution for sustainable agriculture by vermiculture biotechnology using earthworms: Reviving the dreams of Sir Charles Darwin. *Journal of Agricultural Biotechnology and Sustainable Development, 2*, 113–128.
23. Sogbesan, O. A., & Madu, C. T. (2008). Evaluation of earthworm meal as protein feedstuff in diets for Heterobranchus longifilis Valenciennes, 1840 fingerlings under laboratory condition. *Research Journal of Environmental Sciences, 2*, 23–31.
24. Sogbesan, A. O., & Ugwumba, A. A. A. (2008). Nutritional values of some non-conventional animal protein feedstuffs used as fishmeal supplement in aquaculture practices in Nigeria. *Turkish Journal of Fisheries and Aquatic Sciences, 164*, 159–164.
25. Gaviria Gaviria, Y. S., Camaño Echavarría, J. A., & Zapata Montoya, J. E. (2021). Evaluation of the environmental impact of dry chemical silage obtained from the viscera of red tilapia (Oreochromis spp.) using ecological footprint methodology. *Heliyon, 7.*
26. Perez-Martinez, M. M., Noguerol, R., Casales, B. I., Lois, R., & Soto, B. (2018). Evaluation of environmental impact of two ready-to-eat canned meat products using Life Cycle Assessment. *Journal of Food Engineering, 237*, 118–127.
27. Krotscheck, C., & Narodoslawsky, M. (1996). The sustainable process index. A new dimension in ecological evaluation. *Ecological Engineering, 6*, 241–258.

28. Mamouni Limnios, E. A., Ghadouani, A., Schilizzi, S. G. M., & Mazzarol, T. (2009). Giving the consumer the choice: A methodology for Product Ecological Footprint calculation. *Ecological Economics, 68*, 2525–2534.
29. IPCC. (2006). *Directrices Para Los Inventarios Nacionales de Gases de Efecto Invernadero – Elimimación de Desechos Sólidos – Capitulo 3. IPCC.*
30. IPCC. (1996). *Directrices Para Los Inventarios Nacionales de Gases de Efecto Invernadero – Desperdicios – Módulo 6*. vol. 63.
31. IPCC. (2014). *Cambio Climático 2014: Informe de Síntesis. Contribución de los Grupos de trabajo I,II y III al Quinto Informe de Evaluación del Grupo Intergubernamental de Expertos sobre el Cambio Climático.*
32. IDEAM. (2011). *Estimación de Las Emisiones de Dióxido de Carbono Generadas Por Deforestación Durante El Periodo 2005–2010.*
33. AOAC. (2000). *Official methods of analysis of AOAC International*. Association of Official Analysis Chemists International.
34. Osborne, T. B. (1907). *The proteins of the Wheat Kernel*. Carnegie Institution of Washington. https://doi.org/10.5962/bhl.title.22763
35. ICONTEC. (2009). *NTC 4519: Microbiología de Los Alimentos Para Consumo Humano y Animal. Método Horizontal Para El Recuento de Microorganismos.*
36. Garczyńska, M., et al. (2023). Chemical composition of earthworm (Dendrobaena veneta Rosa) biomass is suitable as an alternative protein source. *International Journal of Environmental Research and Public Health, 20.*
37. Rodrigues, M., et al. (2017). Enzymatic hydrolysis of the Eisenia andrei earthworm: Characterization and evaluation of its properties. *Biocatal Biotransformation, 35*, 110–119.
38. Malakahmad, A., Abualqumboz, M. S., Kutty, S. R. M., & Abunama, T. J. (2017). Assessment of carbon footprint emissions and environmental concerns of solid waste treatment and disposal techniques; case study of Malaysia. *Waste Management, 70*, 282–292.
39. Santocki, M., Falniowski, A., & Plytycz, B. (2016). Restoration of experimentally depleted coelomocytes in juvenile and adult composting earthworms Eisenia andrei E. fetida and Dendrobaena veneta. *Applied Soil Ecology, 104*, 163–173.
40. Quiroz Fernández, L. S., Izaquierod Kulich, E., & Menéndez Gutierrez, C. (2018). Estudio del impacto ambiental del vertimiento de aguas residuales sobre la capacidad de autodepuración del río Portoviejo, Ecuador. *Centro Azúcar, 45*, 73–83.
41. Komakech, A. J., Banadda, N. E., Gebresenbet, G., & Vinnerås, B. (2014). Maps of animal urban agriculture in Kampala City. *Agronomy for Sustainable Development, 34*, 493–500.
42. Leinonen, I., Williams, A. G., Wiseman, J., Guy, J., & Kyriazakis, I. (2012). Predicting the environmental impacts of chicken systems in the united kingdom through a life cycle assessment: Egg production systems. *Poultry Science, 91*, 8–25.
43. Singh, A., & Singh, G. S. (2017). Vermicomposting: A sustainable tool for environmental equilibria. *Environmental Quality Management, 27*, 23–40.
44. Komakech, A. J., Zurbrügg, C., Miito, G. J., Wanyama, J., & Vinnerås, B. (2016). Environmental impact from vermicomposting of organic waste in Kampala, Uganda. *Journal of Environmental Management, 181*, 395–402.

Ecological Footprint of Electric Vehicle Charging Infrastructure

Shah Faisal, Dilawar Husain, U. S. Ansari, Govind Rai Goyal, Bhanu Pratap Soni, Fareed Ahmad, Kirti Tewari, and Ravi Prakash

1 Introduction

Sustainable electric transportation has the potential to yield a substantial reduction in greenhouse gas (GHG) emissions associated with the movement of people and goods. Some countries endeavour to attain full electric vehicle (EV) based transportation systems to achieve sustainable development goals. The market share of electric vehicles (EVs) was 28.8% in Norway, 6.4% in the Netherlands, 1.4% in China, and 1% in India as of January 2020 [1]. IEA energy projects estimate that by

S. Faisal · G. R. Goyal
Department of Electrical Engineering, University of Engineering & Management, Jaipur, Rajasthan, India

D. Husain (✉)
Department of Mechanical Engineering, Maulana Mukhtar Ahmad Nadvi Technical Campus, Malegaon, Maharashtra, India

U. S. Ansari
Department of Civil Engineering, S.N.D College of Engineering & Research Centre, Nashik, Maharashtra, India

B. P. Soni
School of Electrical & Electronics Engineering, Fiji National University, Suva, Fiji

F. Ahmad
Department of Electrical Engineering, S.N.D College of Engineering & Research Centre, Nashik, Maharashtra, India

K. Tewari
Department of Mechanical Engineering, National Institute of Technology Sikkim, Ravangla, Sikkim, India

R. Prakash
Department of Mechanical Engineering, Motilal Nehru National Institute of Technology Allahabad, Prayagraj, India

S. S. Muthu (ed.), *Ecological Footprint of Industrial Spaces and Processes*, Environmental Footprints and Eco-design of Products and Processes, https://doi.org/10.1007/978-3-031-69047-1_6

2030 more than 145 million EVs will be travelling roads all over the world [2] and it means a 31% annual compounding growth. The industry's ability to grow more as it will be powered by renewable sources such as the sun and the wind and thereby the carbon emission will be almost zero. In contrast, electricity generation from fossil fuels only increases the charging loads which results in more harm to the environment [3]. For EV users' infrastructural accessibility is possible through a well-thought system which purposely provides accommodation to users in all parts of the highways.

1.1 Charging Types

The EVs are charged using three different charging techniques. All three charging types are depicted in Fig. 1. The following are the specifics of EV charging types: The following are the specifics of EV charging types:

1.1.1 Level 1 Charging

It is the most basic type of EV charging, utilizing a standard household outlet (120 V) to charge an electric vehicle. It offers the slowest charging rate and is best suited for overnight charging or when no other options are available.

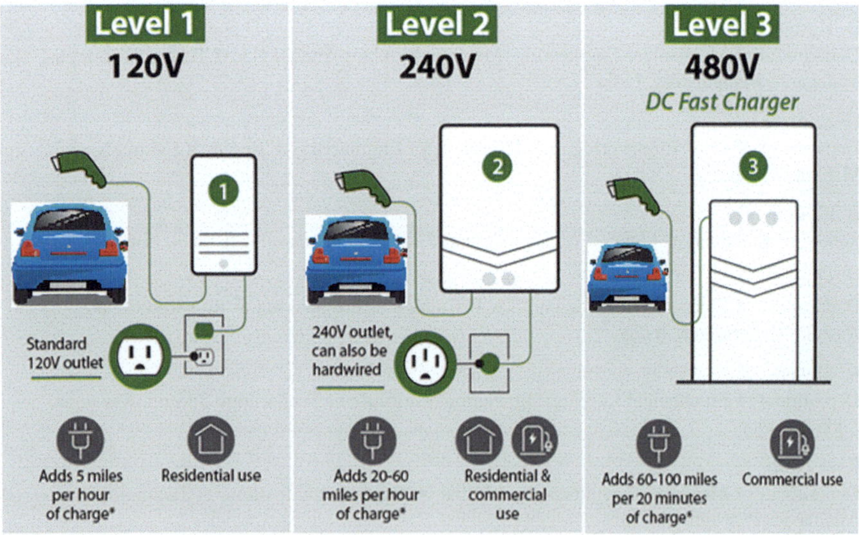

Fig. 1 Different types of EV charging [5]

(i) **Charging Rate:** typically delivers 2–5 miles of range per hour of charging [4].
(ii) **Equipment:** requires only a standard power cord and an adapter for the vehicle.

1.1.2 Level 2 Charging

It is the most common charging option for residential and public charging stations. It uses a higher-powered 240 V outlet, often like those used for large appliances, to provide faster charging compared to Level 1.

(i) **Charging Rate:** typically delivers 10–60 miles of range per hour of charging, depending on the vehicle and charging station capacity [4].
(ii) **Equipment:** requires a dedicated Level 2 charging station that includes a connector compatible with the EV, an electric panel, and appropriate circuitry.

1.1.3 DC Fast Charging (DCFC)

It is a high-power charging option that can rapidly charge an electric vehicle, especially for long-distance travel or in areas with limited charging infrastructure.

(i) **Charging Rate:** it can provide up to 150–350 miles of range in just 20–30 min, depending on the vehicle's capabilities and charging station capacity [4].
(i) **Equipment:** utilizes a specialized charging station with a high-powered DC connection, capable of bypassing the vehicle's onboard charger for more efficient charging.

1.2 Charging Infrastructure

The installation and proliferation of electric vehicle charging stations (EVCS) have become essential components in supporting the mass adoption of EVs. Table 1 presents a comprehensive forecast of EV fast public charging stations spanning from 2015 to 2030. By the year 2030, projections indicate that China will boast 1064 fast charging stations, Europe will host 112, the United States will possess 51, and an additional 58 stations will be distributed across other countries.

The way the installation of EV charging stations is trending today exhibits growing attention toward the electric vehicle infrastructure and its necessity its switch to a sustainable transport system. The data of the forecast illustrate the rate of growth of the fast-charging points in China is from 200 to 1064 by 2030, in Europe by 112, in the USA by 51 and in other countries by 58. These views suggest why there is a necessity to set up charging points to drive infrastructure for the development of the electric vehicle market which is expected to grow in the future.

Table 1 Fast public charging station 2015–2030 unit in thousand [6]

S. No	Year	China	Europe	United States	Other countries
1	2015	12	6	4	6
2	2016	55	9	3	8
3	2017	83	11	3	11
4	2018	111	16	4	14
5	2019	215	25	13	17
6	2020	309	38	17	22
7	2021	470	49	22	28
8	2022	467	51	23	29
9	2023	539	58	26	33
10	2024	611	69	32	37
11	2025	718	74	33	41
12	2026	768	83	37	44
13	2027	842	89	40	48
14	2028	915	97	44	51
15	2029	995	105	47	55
16	2030	1064	112	51	58

1.3 Charging Specifications

EV's suitability for different charging methods may vary, and charging times can be influenced by factors including factors like battery capacity, charging infrastructure, and various other variables. Many studies have been conducted on the circular economy and energy savings, to maximize environmental and social gains, by improving the operations of the electric charging stations with moderate investments [7–9]. The type of energy chosen greatly affects the life cycle assessment of charging stations [10], just the same, it lowers the transport costs [11], increases profit and has the potential for reduction of emissions both at the development stage and in operation [12, 13]. Wang et al. focused on the environmental impact of EV charging when energy is sourced from different production combinations with a call to embrace cleaner electricity production forms [14]. A reduction of 13.4% in greenhouse gas emissions has been observed in EVs by 2020 [15], while Schäfer et al. pointed out that renewable energy is a crucial prerequisite for functioning EV charging stations [16]. All the research done together points to the multidimensional and integral factors critical for fostering sustainable transport options. The various studies based on energy consumption, emission and economics of charging stations are mentioned in Table 2.

The energy demand, emissions and charging cost for EV battery charging can be minimized by using renewable energy systems. The reported battery charging cost is range of 0.1302–0.1430 $/kWh, with a 34.68% reduction in CO_2 emissions compared to conventional grid-based charging stations.

Table 2 Literature review based on energy consumption, emissions & cost of EV charging

Article reference	Energy	Emissions	Cost	Remarks/years
Marinescu et al. [17]			✔	The life cycle cost (LCC) obtained was 584 kUSD (i.e., 1.01 USD for every 1 kUSD) in comparison with 779–1101 kUSD for the renewable-powered charging stations. The cost reductions of the charging process from 53% to 74.9%.
Han et al. [18]	✔		✔	This paper proposes an economic estimation method for PV charging stations using retired EV batteries. The charging service charge of the fast-charging station is equal to 0.1430 USD/kWh.
Shen et al. [19]	✔			DC microgrids are favoured for new energy systems. This paper explores a DC microgrid with PV generation and EV charging, introducing hybrid energy storage (flywheel and battery).
Karmaker et al. [20]	✔	✔		Bangladesh's EV growth strains the energy supply and emits CO_2. The paper proposes a solar and biogas EV charging station to relieve the national grid.
El Sayed et al. [21]		✔	✔	The study created a model that focuses on drone charging stations for aerial goods transportation. Solar energy systems can reduce operational costs and reduce the dependency on gride.
Miralinaghi et al. [22]	✔	✔		Metropolitan authorities want to reduce emissions and shift from fossil fuels to electric transportation. This paper offers a long-term framework to boost electric vehicle adoption.
Mishra et al. [23]		✔		This paper focuses on planning charging station infrastructure for electric vehicles, emphasizing key considerations and ongoing developments with a reduction in emissions.
Gilev et al. [24]		✔	✔	Using microgas turbines (MGT) for electric vehicle charging brings technical and economic advantages. MGTs are smaller than heavy-duty gas turbines and can use various fossil fuels at different pressure levels, creating co-generation opportunities.
Yan et al. [25]	✔		✔	This paper presents a four-stage algorithm for optimizing an EV charging station with solar panels and battery storage. It minimizes operational costs, handles uncertainties, and balances supply and demand through EV charging, grid supply, and load management.

1.4 Indian Scenario

The research examines the ecological footprints of EV charging stations in India to explore their environmental effect and feasibility. The stations that predominantly

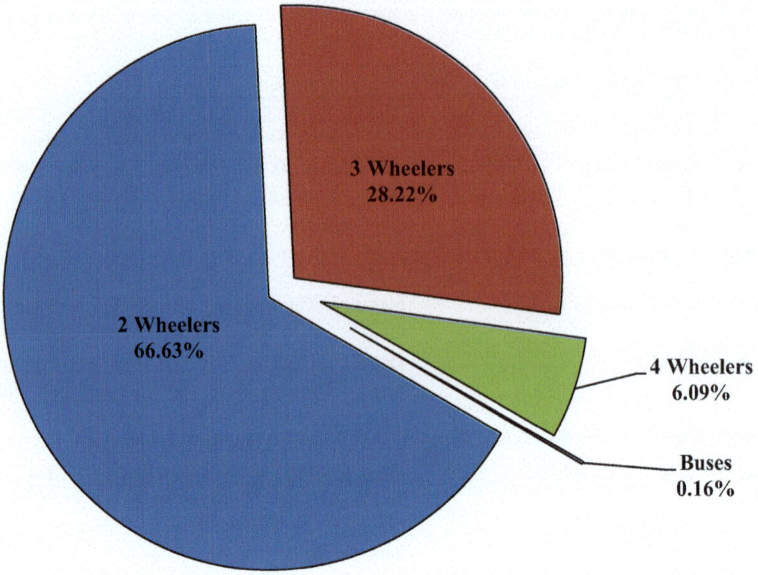

Fig. 2 Market share of EVs in India 2023 [26]

use grid electricity to charge EV batteries are the principal EV chargers, and the ecological footprint is heavily reliant on the source of grid electricity. The market share of EVs in India is depicted in Fig. 2. The research ascertains that the usage of renewable energy technologies such as solar PV and wind empowers battery charging without negative environmental influence. In contrast, relying on the (fossil-borne) conventional power systems to meet EV charging demands will bring about an increase in the ecological footprint because of the emissions from the plants. Therefore, the study campaigns for the replacement of traditional EV charging systems with sustainable charging systems to minimize transportation-related environmental impacts and achieve sustainable development goals.

In various types of EVs composing the Indian market in 2023, electric 2 Wheelers posted the highest share of 66.63%, constituting more than half of EV sales. Among the 3 Wheeler electric vehicles, a share of 28.22% was noted, and the electric 4 Wheelers captured a smaller share of 6.09%. The operating areas of electric buses were 0.16% [26].

2 Materials and Methods

2.1 EV Charging Infrastructure

The DC Fast Charging station featured in the case study is situated along the Solapur – Dhule National Highway, specifically at MIDC Avdhan, Tikkhi, State

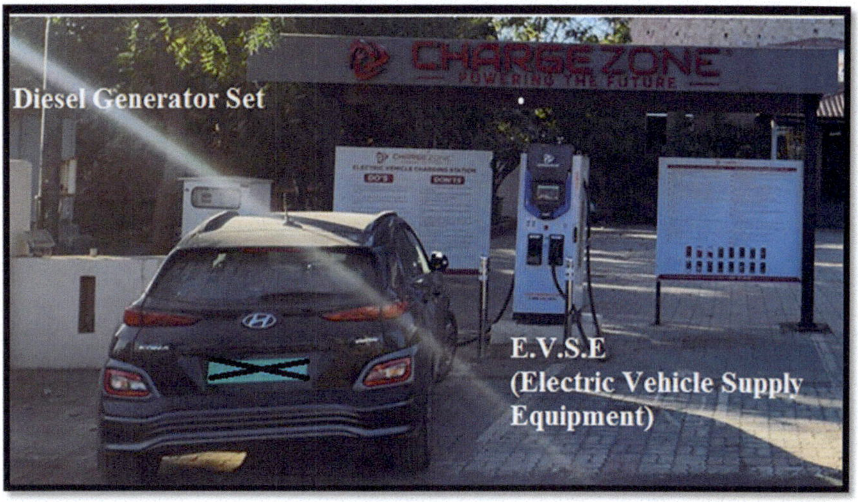

Fig. 3 Exploration of electric vehicle charging infrastructure in case study

Table 3 The details of the case study charging infrastructure [27]

Station name	Charge zone charging station
Power Input/Output Supply	440 V (3 Phase, Alternate Current)/60 kW
Charging Type	DC Fast (Level 3)
Number of Charging Ports	2
Backup Power Source: Diesel Powered Generator Set-up	
Backup Power Capacity	60 kVA (3 Phase, Alternate Current)
Rated Speed	1800 rpm

Maharashtra, India. This station is designed to provide power to electric vehicles via a grid connection, with a diesel power generator serving as a backup/emergency power source. However, concerns arise regarding the environmental impact associated with the diesel generator. Figure 3 illustrates the specifics of the existing EV charging station, which occupies approximately 400 square feet (37.2 m^2) of physical land. The details about the case study EV charging station are mentioned in Table 3.

2.2 Renewable Energy Systems

Although renewable energy systems are considered a clean power source, they still entail energy consumption during various stages such as materials sourcing, transportation, and installation. Consequently, they are responsible for a small amount of greenhouse gas emissions [27]. The government of India is promoting solar PV systems to reach the goal of 100 GW of installed capacity by 2022 [28]. Grid-connected rooftop solar PV systems will play an important role in reaching this target, hence these have been considered in this study.

2.3 Studies on the Ecological Footprint of EV Charging Stations (E_FEV)

The Environmental Impact of EV charging stations hinges on three primary factors: (1) Electricity/fossil fuel consumption, (2) Manpower, and (3) Physical land occupancy. In order to concretize the environmental footprint of electric vehicles' charging infrastructure the combined ecological footprint of these factors is extremely significant. The representation in Fig. 4 shows the flowchart of the Ecological Footprint calculation for EV charging.

(i) **Energy Use (EF_E)**

The Ecological Footprint of EV charging infrastructures is related to the two folds of energy: direct and indirect use. It can be defined by the use of Eq. 1 to compute the EV charging system's EF_E.

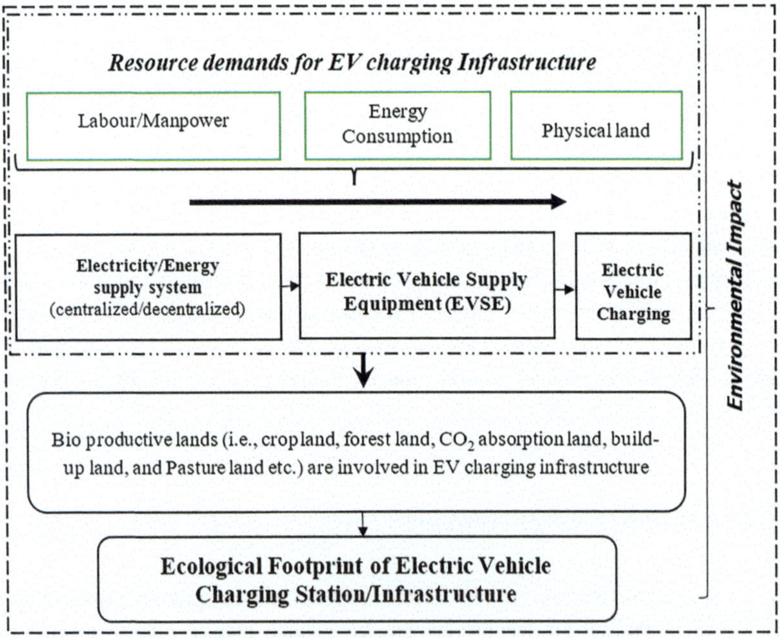

Fig. 4 Methodology for ecological footprint of EV charging infrastructure (self-made)

Table 4 Details of the different parameters used in Eqs. 1, 2, and 3

<table>
<tr><th>Parameters</th><th colspan="3">Details</th></tr>
<tr><td>E_{cF}</td><td colspan="3">Total conventional (electricity, fossil fuel, etc.) energy used by the EV charging station</td></tr>
<tr><td>E_{cR}</td><td colspan="3">Total renewable energy used by the EV charging station</td></tr>
<tr><td>α_{cF}</td><td colspan="3">Emission factor of conventional energy sources (for electricity 0.82 tCO_2/MWh [31], for diesel fuel 3.17 CO_2 kg/kg of diesel [32])</td></tr>
<tr><td>α_{cR}</td><td colspan="3">Emission factor of renewable energy sources</td></tr>
<tr><td>A_f</td><td colspan="3">Average CO_2 absorption capacity of forest (2.7 tCO_2/ha [33])</td></tr>
<tr><td>A_{oc}</td><td colspan="3">Fraction of annual oceanic emission sequestration (i.e., 0.3 [34])</td></tr>
<tr><td>e_i</td><td colspan="3">Equivalence factor of bioproductive land [35]</td></tr>
<tr><td></td><td>Cropland ($e_{cropland}$)
CO_2 absorption land ($e_{CO2\ land}$)
Forest land ($e_{forest\ land}$)
Built-up land ($e_{built\text{-}up\ land}$)
Pasture land ($e_{pasture\ land}$)
Sea productive/marine land ($e_{marine\ land}$)</td><td colspan="2">2.52 gha/ha
1.28 gha/ha
1.28 gha/ha
2.52 gha/ha
0.43 gha/ha
0.35 gha/ha</td></tr>
<tr><td>EF_{food}</td><td colspan="3">Annual Ecological Footprint of food consumption by an Indian (i.e., 0.549 gha/year [36])</td></tr>
<tr><td>M_{rate}</td><td colspan="3">Metabolic calorie burn rate during working</td></tr>
<tr><td></td><td>Activity</td><td>Calories Burn/hour (kcal/h) [6]</td><td>Ecological Footprint (gha/h) [6]</td></tr>
<tr><td></td><td>High activities</td><td>312.2–403.8</td><td>2.09×10^{-4}</td></tr>
<tr><td></td><td>Moderate activities</td><td>203.4–310.6</td><td>1.50×10^{-4}</td></tr>
<tr><td></td><td>Low activities</td><td>110.3–201.9</td><td>0.91×10^{-4}</td></tr>
<tr><td>H_{wh}</td><td colspan="3">Total working hours</td></tr>
<tr><td>$A_{station}$</td><td colspan="3">Total engaged land (hectare) of EV charging stations.</td></tr>
</table>

$$\text{EFE} = \sum (E_{cF}.\alpha_{cF} + E_{cR}.\alpha_{cR}).\left(\frac{1 - A_{oc}}{A_f}\right).e_{CO_{2land}} \tag{1}$$

All the parameters listed in Eq. 1 are mentioned in Table 4.

(ii) **Manpower Impact (EF_M)**

Ecological Footprint of human activities is directly linked with the metabolic calorie rate burn [29, 30]. Table 4 lists some exercise metabolic burn rates in the calorie burn parameters. The Ecological Footprint of workforce/manpower for EV charging stations is Eq. 2:

$$\text{EFM} = \frac{EF_{food}}{365\ x\ Daily\ Calories\ Burn} \times \sum (M_{rate}.H_{wh})_i \tag{2}$$

All the parameters listed in Eq. 2 are mentioned in Table 4.

(iii) **Physical land (EF_L)**

The environmental impact of physical land related to EV charging infrastructure/ station is estimated by Eq. 3:

$$EF_L = A_{station} \times e_{built-up\ land} \tag{3}$$

All the parameters listed in Eq. 3 are mentioned in Table 4.

3 Results and Discussions

The yearly electricity usage at the EV charging station amounts to 174,649 kWh. Of this total, approximately 98% comes from the grid, with the remaining 2% generated by a diesel generator set.

3.1 *Ecological Footprint of EV Charging Station with Grid Electricity*

The current ecological impact of the EV charging infrastructure is estimated to be equivalent to 48.52 global hectares. Energy consumption within the EV charging station significantly drives this impact, accounting for 98% of the station's total yearly ecological footprint. The assessment of the EV charging infrastructure's ecological footprint includes the following details.

(i) **Energy Consumption (EF_E)**

According to reports, the ecological footprint of India's national grid is 0.000272 gha/kWh [37]. 46.44 gha is the estimated overall ecological footprint associated with grid electricity consumption. According to the literature, the ecological footprint of diesel fuel is 0.00158 gha/liter [28]. The total annual electricity consumption in the EV charging infrastructure, total EV charging and PV electricity generation are shown in Fig. 5. The estimated ecological footprint of using power (produced by a diesel generator set) is 1.27 gha. As a result, for the evaluation year 2023, the predicted Ecological Footprint of the entire electricity consumption in the case study EV charging is 47.71 gha.

(ii) **Labour**

The charging infrastructure operates 24 × 7 throughout the year and one operator is required to handle the EV charging process. Therefore, the Full Time Equivalent of the charging station is about 3. The ecological footprint associated with human labour is documented at 0.000091 global hectares per hour. The total ecological footprint attributed to labour is computed to be 0.80 global hectares.

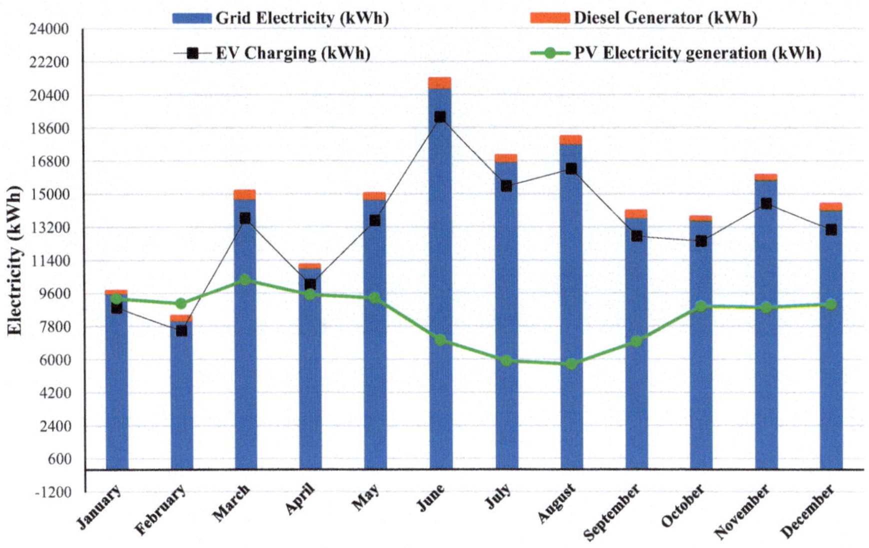

Fig. 5 The total annual electricity consumption in the EV charging infrastructure, total EV charging and PV electricity generation

(iii) **Land**

The EV charging station utilized 0.0037 hectares (equivalent to 400 square feet) of land space. The ecological footprint linked with land usage is approximated to be 0.01 global hectares.

In the assessment year of 2023, the EV charging station received a total electricity supply of 174,649 kWh. However, due to factors like the 90% efficiency of the EVSE and miscellaneous electricity consumption within the infrastructure (such as lighting, water cooler, etc.), only 157,200 kWh of energy was utilized to charge EV batteries. The ecological footprint associated with EV charging is approximately 3.1×10^{-4} global hectares per kilowatt-hour of battery charging.

3.2 *Ecological Footprint of EV Charging Station with Solar PV System*

Introducing a renewable power system is the starting point for an operation that will be both friendly to the environment and cost-effective. Using solar energy conversion with the use of photovoltaic (PV) panels the charging stations can be relieved from environmentally unfriendly energy sources, drastically reducing their ecological footprint. Figure 6 illustrates the various energy sources connected to the EV charging station and indicates the quantity of electricity transmitted from EVSE (Electric Vehicle Supply Equipment) to batteries.

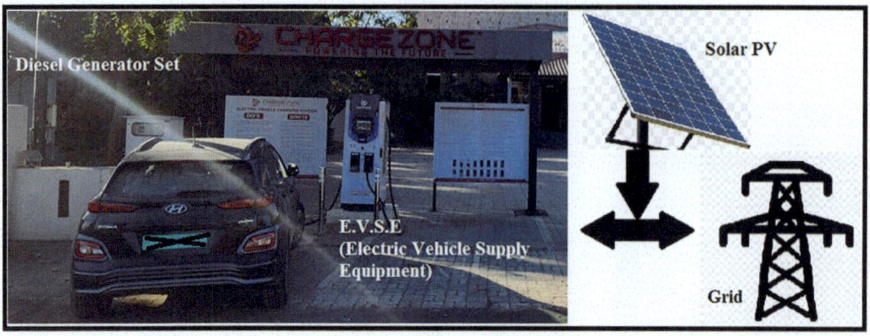

Fig. 6 Interconnection of EV charging stations with solar photovoltaic (self design)

The RETScreen software program is applied to estimate that the planned solar photovoltaic plant with a capacity of 63 kW would be able to provide 57.3% of the total annual energy consumption required energy for the charging station in the year 2023. The greenhouse gas emission related to the grid-coupled solar PV system is saved at 1.59×10^{-5} gha/kWh. For the envisioned system, it is predicted that the footprint of the ecosystem is going to be about 0.001 global hectares Given the above solar PV system, this EV charging infrastructure's ecological footprint is approximately 22.9 gha. On the contrary, the measure of the ecological footprint of the charging system of EVs together with the installation of the proposed grid-connected solar PV system is approximately 1.45×10^{-4} gha/kWh of charging the battery. Based on the fact that a grid-tied solar generation could virtually reduce greenhouse impact up to 47.2%, it can be seen that smart grid and storage are capable of increased contribution to the greenhouse gases. Tables 5 present the details of the proposed solar photovoltaic system.

3.3 *Economic Assessment*

The price of the EV charging station set up is estimated at 49,491.52$, with the average yearly operation cost of around 31,683.82$, taking into account a kWh of electricity grid at $0.133. The EV charging station, the LCC of which is estimated to be $444,271.91, is to be constructed. The total cost of battery charging for the vehicle lifetime is assessed at $0.169 per kWh (equivalent to R.s. 13.87 per kWh). Incorporating the proposed hybrid system increases the total installation cost of the EV charging station that amounts to $126,283.70, with the installation cost the solar photovoltaic set at $853.24 per one kilowatt in India. Therefore, the LCC of the EV charging station that is planned to have the hybrid system is projected at $282,626.92. Substitution of the electricity for the proposed solar PV system, the charging cost of battery per kWh decreased to $0.107 ($9.18 Indian Rupee) which

Table 5 Estimated capacity of the proposed solar photovoltaic system

S. No	Parameters	Unit	Value
Solar Module			
1	Manufacturer	–	Trina Solar
2	System Power Capacity	kW	63
3	Module Type	–	Poly-Si
4	Model	–	poly-Si – TSM – PD14/340 W [38]
5	Panel Efficiency	%	17.49 [38]
6	Number of panels	–	185
7	Solar Collector Area	m^2	360
8	Temperature Coefficient	%/ °C	0.4
9	Bifacial call adjustment factor	%	5
10	Miscellaneous Losses	%	5
Inverter			
12	Efficiency	%	95
13	Capacity	kW	65
14	Miscellaneous Losses	%	5
System Performance			
15	Capacity Factor	%	18.2

might significantly increase the EV charging annuity income by up to $19,576.64. What is more, the proposed solar PV system can help the rate of battery charging to be reduced by as much as 36.7% which would lead to the reduction of up to the current charging costs.

Figure 7 shows the environmental impact (Ecological Footprint) of EV charging over the current charging station and the cost investigation of using the alternative hybrid system. The proposed hybrid system demonstrates attractive perspectives of saving energy of batteries' charging up to 90.1% of that of the present charging station. Lastly, it may be a cost saver, with savings as high as 36.7% on battery charging charges compared to present expenditures. Such implication suggests that EV charging stations have a pivotal role in limiting the environmental impact of EV charging which lines up the move towards sustainable development.

4 Conclusions

One of the issues the completion of electric vehicle charging infrastructure has to be an intermediate study that brings the perspective of both the environment and the economics. For instance, one EV charging station in the scale is rated year by 2023 and globally by hectares in 48.52. In contrast to the global footprint per kilowatt hour of battery-charging this is backgrounded at 3.1×10^{-4} global hectares. Nevertheless, a solar PV system can be used to mitigate the environmental footprint of a battery

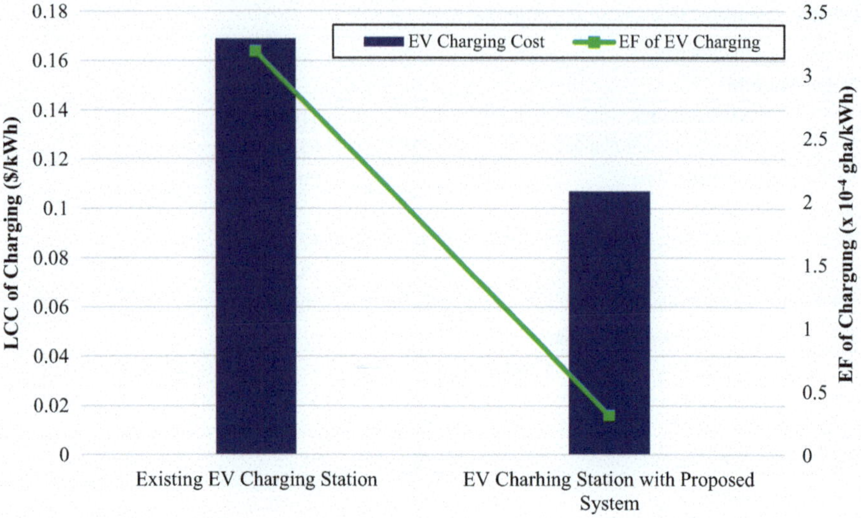

Fig. 7 Ecological footprint and cost of EV charging

charging process to 1.45×10^{-4} global hectares per kWh meanwhile the reduction potential reaching to 47.2% compared with the case of the existing scheme.

Most important is the development of renewable energy that will be used in charging EV stations to minimize the environmental issues related to charging. The use of grid-scale connections for the integration of sustainable energy sources such as solar, wind, and geothermal into the overall power system design will offer the opportunity to significantly reduce greenhouse gas emissions and divert from fossil fuel use.

Carrying out the economic assessment of EV charging station establishments calls for the analysis of all their financial aspects. To maintain the longevity of such stations, it is crucial to achieve a balance between initial investment outlays, operational costs and revenue sources. Providing careful financial evaluation in the long run and coming up with some new business models is necessary for governments, businesses and investors to ensure the profitability of the investment. Governments act as a catalytic enzyme in improving the ecological and economic profitability of EV charging stations. They can subsidize the installation and operation of charger stations using benchmarks, tax credits, and other money-saving measures. Enhancing the permitting procedures and building a less complicated regulatory system that is supportive of standards, policies, and regulations will also boost the amount of private and public investment in EV charging infrastructure.

The recycling criteria for the solar panels were overlooked in the study. However, the advantage of the proposed solar PV system is that it not only cuts down the ecological footprint and charging fee for EVs but also should be highly commended. Low-impact charging systems are based on two main criteria: their effectiveness and scalability.

References

1. Meyer, D., & Wang, J. (2018). Integrating ultra-fast-charging stations within the power grids of smart cities: A review. *IET Smart Grid, 1*(1), 3–10.
2. International Energy Agency (IEA) report: https://www.iea.org/reports/global-ev-outlook-2021
3. Ajanovic, A. (2014). Promoting environmentally benign electric vehicles. *Energy Procedia, 57*, 807–816.
4. ZDWL, "Proven EV charger manufacturer". https://zdwl-tec.com/news/ev-charger-levels/
5. Value Electric: Quality affordable electrical services. https://www.readysetgoev.com/articles/level-1-vs-level-2-ev-chargers-finding-your-perfect-match
6. Faisal, S., Soni, B. P., Goyal, G. R., Bakhsh, F. I., Husain, D., & Ahmad, A. (2023). Reducing the ecological footprint and charging cost of electric vehicle charging station using renewable energy based power system. *e-Prime-Advances in Electrical Engineering, Electronics and Energy*, 100398.
7. Ma, Z., Wang, J., & Ma, X. (2021). A review of smart charging infrastructure for electric vehicles: Current state, future trends, and challenges. *Journal of Cleaner Production, 293*, 126129.
8. Sioshansi, R., Miller, C., & Marano, V. (2020). Economic analysis of public electric vehicle charging infrastructure deployment. *Transportation Research Part D: Transport and Environment, 81*, 102247.
9. Díaz-González, M., Gerbec, M., Varela, M., & Sumper, A. (2021). Electric vehicle charging stations integrated with renewable energy sources and energy storage systems. *Energies, 14*(4), 1029.
10. Hoang, A., Börjesson, M., & Vassileva, I. (2019). Life cycle assessment of public charging infrastructure for electric vehicles: A case study in Sweden. *Journal of Cleaner Production, 211*, 420–429.
11. Zhan, D., Su, J., Zhang, L., Guo, S., & Wang, C. (2022). A review on dynamic pricing for electric vehicle charging: Models, methods, and future trends. *Energy Reports, 8*, 648–662.
12. Gopalakrishnan, V., Sivakumar, A., & Kandlikar, M. (2018). Policy mechanisms to promote private investment in electric vehicle charging infrastructure. *Applied Energy, 229*, 1267–1280.
13. Dong, G., Ma, J., Wei, R., & Haycox, J. (2019). Electric vehicle charging point placement optimisation by exploiting spatial statistics and maximal coverage location models. *Transportation Research Part D: Transport and Environment, 67*, 77–88.
14. Wang, L., Nian, V., Li, H., & Yuan, J. (2021). Impacts of electric vehicle deployment on the electricity sector in a highly urbanised environment. *Journal of Cleaner Production, 295*, 126386.
15. Wu, Z., Wang, M., Zheng, J., Sun, X., Zhao, M., & Wang, X. (2018). Life cycle greenhouse gas emission reduction potential of battery electric vehicle. *Journal of Cleaner Production, 190*, 462–470.
16. Schäfer, A. W., Kempton, W., & Boies, A. (2017). Integration of renewable energy sources in electric vehicle charging stations. *Applied Energy, 188*, 149–158.
17. Marinescu, C. (2022). Progress in the development and implementation of residential EV charging stations based on renewable energy sources. *Energies, 16*(1), 179.
18. Han, X., et al. (2018). Economic evaluation of a PV combined energy storage charging station based on cost estimation of second-use batteries. *Energy, 165*, 326–339.
19. Shen, L., et al. (2020). Hierarchical control of DC micro-grid for photovoltaic EV charging station based on flywheel and battery energy storage system. *Electric Power Systems Research, 179*, 106079.
20. Karmaker, A. K., et al. (2018). Feasibility assessment & design of hybrid renewable energy based electric vehicle charging station in Bangladesh. *Sustainable Cities and Society, 39*, 189–202.

21. ElSayed, M., Foda, A., & Mohamed, M. (2022). Autonomous drone charging station planning through solar energy harnessing for zero-emission operations. *Sustainable Cities and Society, 86*, 104122.
22. Miralinaghi, M., et al. (2020). Designing a network of electric charging stations to mitigate vehicle emissions. In *2020 Forum on integrated and sustainable transportation systems (FISTS)*. IEEE.
23. Mishra, S., et al. (2021). A comprehensive review on developments in electric vehicle charging station infrastructure and present scenario of India. *Sustainability, 13*(4), 2396.
24. Gilev, B., et al. (2022). Modeling and simulation of a low-cost fast charging station based on a micro gas turbine and a supercapacitor. *Energies, 15*(21), 8020.
25. Yan, Q., Zhang, B., & Kezunovic, M. (2018). Optimized operational cost reduction for an EV charging station integrated with battery energy storage and PV generation. *IEEE Transactions on Smart Grid, 10*(2), 2096–2106.
26. India's EV sales surge to 1.5 million in 2023, up 50%. https://energy.economictimes.indiatimes.com/news/power/indias-ev-sales-surge-to-1-5-million-in-2023-up-50-report/106971357
27. CHARGEZONE: Powering the future. https://chargezone.com. Accessed 28 Dec 2022.
28. Biswas, A., Husain, D., & Prakash, R. (2021). Life-cycle ecological footprint assessment of grid-connected rooftop solar PV system. *International Journal of Sustainable Engineering, 14*(3), 529–538.
29. Ministry of New and Renewable Energy (MNRE), Government of India. (2018). *India Release State Targets for 40 GW Rooftop Solar by 2022*. http://www.pvtech.org/news/india_releases_individual_state_targets_for_40gw_rooftop_solar_by_2022
30. Husain, D., & Prakash, R. (2019). Life cycle ecological footprint assessment of an academic building. *Journal of The Institution of Engineers (India): Series A, 100*, 97–110.
31. Martínez-Rocamora, A., Solís-Guzmán, J., & Marrero, M. (2017). Ecological footprint of the use and maintenance phase of buildings: Maintenance tasks and final results. *Energy and Buildings, 155*, 339–351.
32. Ministry of Power Central Electricity Authority (MPCEA), "Government of India. (2016). *CO_2 Baseline Database for the Indian Power Sector, User Guide, 2016*. http://www.cea.nic.in/reports/others/thermal/tpece/cdm_co2/user_guide_ver11.pdf. Accessed on Dec 2022.
33. European Environment Agency (EEA). (2013). *EMEP/EEA air pollutant emission inventory guidebook – 2013*. http://www.eea.europa.eu/publications/emep-eea-guidebook-2013
34. Mancini, M. S., Galli, A., Niccolucci, V., Lin, D., Bastianoni, S., Wackernagel, M., & Marchettini, N. (2016). Ecological Footprint: Refining the carbon Footprint calculation. *Ecological Indicator, 61*(2), 390–403.
35. Scripps Institution of Oceanography (SIO), The Keeling Curve (2017). https://scripps.ucsd.edu/programs/keelingcurve/2013/07/03/how-much-co2-can-the-oceans-take-up/. Accessed on Apr 2022.
36. Global Footprint Network (GFN). (2016). *Working Guidebook to the National Footprint Accounts: 2016*. https://www.footprintnetwork.org/content/documents/National_Footprint_Accounts_2016_Guidebook.pdf
37. Ansari, Y., Husain, D., Haadi, S. M., Haloi, J., & Prakash, R. (2022). Life cycle ecological footprint of building: A case study of low-rise tropical residential building. *International Journal of Environmental Science and Technology, 20*, 1–16.
38. Husain, D., Ahmad, A., Sharma, M., Tewari, K., & Prakash, R. (2022). Ecological footprint of jatropha biodiesel production at low scale. In *Environment & energy, environmental science, engineering and technology, environmental sciences*. Newly Published Books, Nova, Science and Technology, Special Topics. ISBN: 978-1-68507-548-4. https://doi.org/10.52305/KJOY6010

Index

S. S. Muthu (ed.), *Ecological Footprint of Industrial Spaces and Processes*, Environmental Footprints and Eco-design of Products and Processes, https://doi.org/10.1007/978-3-031-69047-1

The manufacturer's authorised representative in the EU is Springer Nature Customer Service Centre GmbH, Europaplatz 3, 69115 Heidelberg, Germany. If you have any concerns regarding our products, please contact ProductSafety@springernature.com

Printed and bound by CPI Group (UK) Ltd, Croydon, CR0 4YY
01/10/2026
02225980-0002